A TEXT BOOK OF

ENGINEERING CHEMISTRY-II

(BASIC CHEMISTRY)

SEMESTER – II
FIRST YEAR DIPLOMA COURSES IN ENGINEERING AND TECHNOLOGY

AS PER SBTE'S NEW REVISED SYLLABUS
FOR JHARKHAND, JULY 2017

S. N. NARKHEDE

M.Sc., M.I.S.T.E.
Retired, Head of Applied Chemistry Department,
Government Polytechnic, Sadar,
Nagpur (M.S.)

Mrs. R. M. GANGRADE

M.Sc.
V.B.V. Polytechnic,
Vasai,
Mumbai

F.Y. DIPLOMA (SEMESTER – II) ENGINEERING CHEMISTRY - II　　　　ISBN 978-93-87686-22-9

Second Edition　:　January 2019

©　　　　　　　　:　**Authors**

Published By :
NIRALI PRAKASHAN
Abhyudaya Pragati, 1312, Shivaji Nagar,
Off J.M. Road, PUNE – 411005
Tel - (020) 25512336/37/39, Fax - (020) 25511379
Email : niralipune@pragationline.com

➢ DISTRIBUTION CENTRES

PUNE

Nirali Prakashan　:　119, Budhwar Peth, Jogeshwari Mandir Lane, Pune 411002, Maharashtra

(For orders within Pune)　Tel : (020) 2445 2044, 66022708, Fax : (020) 2445 1538; Mobile : 9657703145

Email : niralilocal@pragationline.com

Nirali Prakashan　:　S. No. 28/27, Dhayari, Near Asian College Pune 411041

(For orders outside Pune)　Tel : (020) 24690204 Fax : (020) 24690316; Mobile : 9657703143

Email : bookorder@pragationline.com

MUMBAI

Nirali Prakashan　:　385, S.V.P. Road, Rasdhara Co-op. Hsg. Society Ltd.,

Girgaum, Mumbai 400004, Maharashtra; Mobile : 9320129587

Tel : (022) 2385 6339 / 2386 9976, Fax : (022) 2386 9976

Email : niralimumbai@pragationline.com

➢ DISTRIBUTION BRANCHES

JALGAON

Nirali Prakashan　:　34, V. V. Golani Market, Navi Peth, Jalgaon 425001, Maharashtra,

Tel : (0257) 222 0395, Mob : 94234 91860; Email : niralijalgaon@pragationline.com

KOLHAPUR

Nirali Prakashan　:　New Mahadvar Road, Kedar Plaza, 1st Floor Opp. IDBI Bank, Kolhapur 416 012

Maharashtra. Mob : 9850046155; Email : niralikolhapur@pragationline.com

NAGPUR

Nirali Prakashan　:　Above Maratha Mandir, Shop No. 3, First Floor,

Rani Jhanshi Square, Sitabuldi, Nagpur 440012, Maharashtra

Tel : (0712) 254 7129; Email : niralinagpur@pragationline.com

DELHI

Nirali Prakashan　:　4593/15, Basement, Agarwal Lane, Ansari Road, Daryaganj

Near Times of India Building, New Delhi 110002 Mob : 08505972553

Email : niralidelhi@pragationline.com

BENGALURU

Nirali Prakashan　:　Maitri Ground Floor, Jaya Apartments, No. 99, 6th Cross, 6th Main,

Malleswaram, Bengaluru 560003, Karnataka; Mob : 9449043034

Email: niralibangalore@pragationline.com

Other Branches : Hyderabad, Chennai

niralipune@pragationline.com　|　www.pragationline.com

Also find us on 🇫 www.facebook.com/niralibooks

Preface

It gives us a sense of satisfaction to bring out this New revised edition of **Engineering Chemistry - II**.

This book aims at providing a complete coverage of the needs of First Year students as per S.B.T.E's. revised syllabus. The entire revised syllabus has been covered keeping in view the non-availability of the complete subject matter through a single source.

The difficult articles have been explained in a simple language providing, wherever necessary, neat and well explained diagrams so that even an average student may be able to follow it independently. A sufficient number of solved examples and problems with answers are given at the end of each topic.

We are grateful to the publisher Shri. Dineshbhai Furia, Shri. Jignesh Furia and staff of Nirali Prakashan, especially Mr. Santosh Bare, Mr. Kiran Velankar and Mrs. Anjali Muley Pune for providing all facilities required and bringing out this book in the short span of time at their disposal.

We sincerely hope this edition will be well received by students and teachers of all polytechnics.

The authors will appreciate suggestions from teachers and students for the improvement of the book.

We take this opportunity to extend our good wishes to the students and teachers of all polytechnics.

– Authors

Syllabus

1. ELECTROCHEMISTRY   **(Hours 04, Marks 08)**

Conductivity of electrolytes – Concepts of Ohm's law, Specific conductance, Specific resistance, Equivalent conductivity and Molar conductance, Variation of specific, molar and equivalent conductance with dilution. Concept of : Cell constant, pH, pOH and Buffer solution. Numericals based on pH and pOH. Applications of pH and buffer solution.

2. METALS AND ALLOYS   **(Hours 12, Marks 24)**

2.1 Metals :

Definition of metallurgy, Brief introduction of the terms involved in metallurgy.

Metallurgy of Iron :

Resources of Iron, Important ores of iron, Extraction process, Smelting in blast furnace, Chemical reactions in blast furnace. Composition of pig iron. Engineering applications of Pig iron, Cast iron, Wrought iron or Malleable iron.

Metallurgy of Copper :

Important ores of copper, Extraction of copper from chief ore. Engineering properties of copper and applications.

Metallurgy of Aluminium :

Important ores of aluminium, Extraction of aluminium from alumina by electrolytic reduction process, Electrolytic refining of aluminium, Engineering properties of aluminium and uses.

2.2 Alloys :

Ferrous Alloys :

Various methods of steel making, Composition, properties and applications of plain carbon steel (low carbon, medium carbon, high carbon and very hard steel) and effect of various alloying elements (Cr, W, V, Ni, Mn, Mo, Si) etc. on steel.

Non-Ferrous Alloys :

Copper alloys : Brass, Bronze, Nickel, Silver or German silver, their composition, properties and applications.

Aluminium alloys : Duralumin, Magnalium, their composition, properties and applications.

Other Alloys : Definition, composition, properties and applications of Soft solder, Tinmann's solder, Brazing alloys, Plumber's solder, Rose metal.

3. NON-METALLIC ENGINEERING MATERIALS   **(Hours 06, Marks 12)**

3.1 Ceramics :

Definition, Properties and Engineering applications. Types - Structural ceramics, Facing material, Refractories, Fine ceramics, Special ceramics.

3.2 Refractories :

Definition, Properties, Applications and Uses of Fire clay bricks, Silica bricks and Masonry bricks.

3.3 **Composite Materials :**

Definition, Properties, Advantages, Applications and Examples.

3.4 **Adhesives :**

Definition, Characteristics, Advantages of adhesives, Examples such as Phenol-Formaldehyde resin, Urea-Formaldehyde resin, Epoxy resin – their properties and applications as an adhesive.

4. **WATER** (Hours 5, Marks 10)

Characteristics, Sources, Impurities, Hard and soft water, Causes of hardness, Types of hardness, Degree of hardness, Boiler and steam generation, Scale and sludge formation – Causes, Disadvantages, Softening methods such as Boiling, Clark's, Soda ash, Lime soda, Zeolite and Ion exchange methods with principle, chemical reactions. Plumbo solvency and its removal. Numerical problems.

5. **CORROSION** (Hours 09, Marks 16)

Definition of corrosion, Types of corrosion (Dry and Wet chemical corrosion) and their mechanism. Protection of metal from corrosion (corrosion control). Applications of protective coatings like metal coating such as Galvanising, Tinning, Metal spraying, Sherardizing, Electroplating and Metal cladding.

PAINTS AND VARNISHES :

Paints :

Definition, Characteristics of a good paint, Constituents and their functions and examples, Methods of applications. Introduction to Chemical resistant paints, Heat resistant paint, Cellulose paint, Luminous paints, Emulsion paints, Metal paints, Cement paints, Water paint or Distempers.

Varnishes :

Definition, Characteristics, Constituents, Types, Composition, Properties and Applications of Japans, Enamels, Lacquers.

6. **LUBRICANT AND LUBRICATION** (Hours 06, Marks 10)

Lubricant – Definition, Classification with examples.

Functions of lubricant, Lubrication – Mechanism of lubrication (fluid film, boundary and extreme pressure). Physical characteristics of lubricants such as Viscosity, Viscosity index, Oiliness, Volatility, Flash and fire point, and Cloud and pour point. Chemical characteristics such as Acid value or Neutralization number, Emulsification, Saponification value, Selection of lubricants, Characteristics of Transformer oil.

□□□

Contents

□□□

ELECTROCHEMISTRY

1.1 CONDUCTIVITY OF ELECTROLYTES

- The substances which allow the flow of electricity through them are called conductors. The flow of electricity involves the flow of electrons. On the basis of the mechanism of transfer of electrons, conductors are classified as electronic and electrolytic conductors.

Electronic conductors: The conductors in which the conduction of electricity takes place by a direct flow of electrons under the influence of applied potential are called electronic conductors. For example, solid and molten metals like Cu, Al.

Electrolytic conductors: The conductors in which the conduction takes place by the migration of positive and negative ions are called electrolytic conductors.

For example, solutions of ionic solids, strong and weak acids and bases.

Difference between electronic and electrolytic conductors:

Electronic conductors	Electrolytic conductors
(i) Conduction of electricity occurs by flow of electrons through the conductor.	(i) The electron transfer occurs by migration of positive and negative ions towards the electrodes.
(ii) The conduction process does not involve chemical changes and the transfer of matter.	(ii) The conduction process involves chemical changes and the transfer of matter.
(iii) The resistance of the conductor increases and conductivity decreases with increasing temperature.	(iii) The resistance decreases and conductivity increases with rise in temperature.

The nature of solutions can be determined by measuring their conductivities.

The substances whose aqueous solution conduct the electricity are called electrolytes whereas those whose aqueous solution conduct the electricity to a very small extent are non-electrolytes. The conductivity of solutions of electrolyte is much higher than that of water whereas in case of non-electrolytes it is same as that of water.

Electrolytes: KCl, acetic acid, NaCl.

Non-electrolytes: Urea, glucose.

Depending on the extent of dissociation, electrolytes are classified as strong and weak electrolytes.

Strong electrolytes: The substances that are almost completely ionized in aqueous solutions are called strong electrolytes. e.g. strong acids, strong bases, ionic salts.

Weak electrolytes: The substances that are ionized to a very small extent in aqueous solutions are weak e.g. weak acids and bases.

The conductivity of aqueous solution of strong electrolytes is greater than that of aqueous solution of weak electrolytes with the same concentration.

The mobility of electrons through the metallic lattice of metals make them good conductors of electricity. Metals offer some resistance to the flow of an electric current. Solutions of electrolytes have a much higher resistance than metallic conductors. So, the electrolytic conduction depends upon the free movement of ions through solution. The factors which affect the electrical conductivity of solutions of electrolytes are

1. the inter-ionic attraction
2. the solvation of ion
3. the viscosity of the solvent.

1.2 CONCEPT OF OHM'S LAW

- **Ohm's law - Statement :** Physical state of a conductor (material, length, area and temperature) remaining the same, the electric current flowing through a conductor is directly proportional to the potential difference across it.

$$I \propto V$$

i.e.
$$V \propto I$$

$$V = \text{Constant} \times I$$

$$\frac{V}{I} = \text{Constant} = R$$

OR

- **Ohm's law - Statement :** When the physical state of a metallic conductor remains constant, the electric current through it is directly proportional to the potential difference between the ends of a conductor.

- The physical state means length, area, temperature, etc.

OR

- **Ohm's law – Statement :** Physical state of a conductor (material, length, area and temperature) remaining the same, the ratio of potential difference (V) to current (I) is always constant.

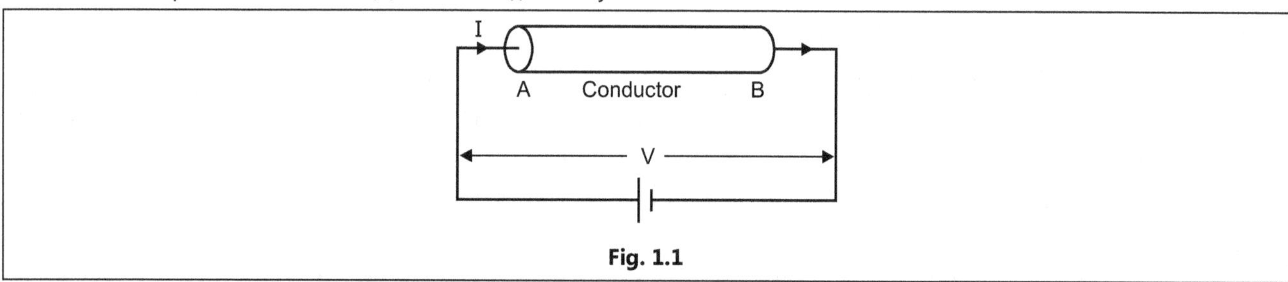

Fig. 1.1

$$V \propto I$$

$$V = \text{Constant} \times I$$

$$\frac{V}{I} = \text{Constant}$$

$$\boxed{\frac{V}{I} = R}$$

or $$\boxed{V = IR}$$

1.3 SPECIFIC RESISTANCE

- **Resistance :** The resistance R of a conductor is
  (i) directly proportional to its length and
  (ii) inversely proportional to its area of cross-section.

Hence,
$$R \propto \frac{l}{a}$$

i.e.
$$R = \rho \frac{l}{a} \qquad \qquad \dots (1.1)$$

where 'ρ' is a constant depending on the nature of the material of conductor, and is called the *'specific resistance'* or *'resistivity'*. If $l = 1$ cm and $a = 1$ sq. cm, then $\rho = R$ ohms.

- Hence, specific resistance may be defined as,

  "The resistance of a uniform column of the material of the conductor having a length of 1 cm and a cross-section of 1 sq. cm."

  Units : As we know that,

$$\rho = R \cdot \frac{a}{l} = \frac{\text{ohm (cm}^2)}{\text{cm}} = \text{ohm. cm.} \qquad \qquad \dots (1.2)$$

1.4 SPECIFIC CONDUCTIVITY (κ)

- The specific conductance of a conductor is the reciprocal of specific resistance, and is generally denoted by κ (Greek, small kappa). Equation (1.2) can be written as :

$$R = \frac{1}{\kappa} \cdot \frac{l}{a} \text{ ohms} \qquad\qquad \left[\because \rho = \frac{1}{\kappa} \right]$$

$$\kappa = \frac{1}{R} \cdot \frac{l}{a} \text{ ohm}^{-1}.\text{ cm}^{-1} \qquad\qquad \ldots (1.3)$$

If l = 1, and a = 1 sq. cm. then the equation (1.3) becomes as

$$\kappa = \frac{1}{R} = \rho$$

or $\qquad\qquad \kappa = \rho$

- Thus, the specific conductance (κ) of a conductor is defined as *'the conductivity offered by a solution of length 1 cm and area of unit cross section or specific conductivity (κ) is the conductance of a one centimeter cube of the substance or solution.'*

Units : The unit of specific conductance is ohm^{-1} cm^{-1} or mhos/cm.

S.I. System : (It is the unit named as Sir. W. Siemens a eminent electrical engineer.)

In SI system, unit of specific conductivity is S.m^{-1} (where S is Siemens, which is the unit for conductance i.e. ohm^{-1} or mho).

Since, $\kappa = \dfrac{1}{\rho} = \dfrac{l}{Ra} = \dfrac{1}{R} \times \dfrac{l}{a} = \dfrac{1}{\text{ohm}} \times \dfrac{\text{cm}}{\text{cm}^2} = \text{ohm}^{-1}\ \text{cm}^{-1}$

The greater the specific conductance, the more readily does the electrolyte allow the passage of an electric current through it. The specific conductance of an electrolyte depends not only upon its nature, but also upon the concentration of solution.

1.5 EQUIVALENT CONDUCTIVITY (λ_v)

- It is defined as the conductivity of a solution containing 1 gm equivalent of the solute/electrolyte, when placed between two sufficiently large electrodes, which are placed 1 cm apart. It is denoted by (lambda), λ_v, where 'V' ml is the volume of the solution containing 1 gm equivalent of electrolyte/solute when placed between the electrodes.

$$\therefore \qquad \text{Equivalent conductivity } (\lambda_v) = \left\{ \begin{array}{c} \text{Volume (in cm}^3) \\ \text{containing 1 gm} \\ \text{equivalent of} \\ \text{solute} \end{array} \right\} \times \left\{ \begin{array}{c} \text{Specific conductivity} \\ \text{of solution} \\ \text{i.e. kappa} \end{array} \right\}$$

or $\qquad\qquad \lambda_v = V \times k$

Therefore, equivalent conductivity can also be defined as, *the product of specific conductivity and the volume (V ml) of the solution containing 1 gm equivalent of the solute.*

If 'c' is the concentration in gm equivalent per litre, then,

$$\lambda_v = \frac{1000\ k}{c} \qquad\qquad \ldots (1.4)$$

It is measured in reciprocal ohm or mho.

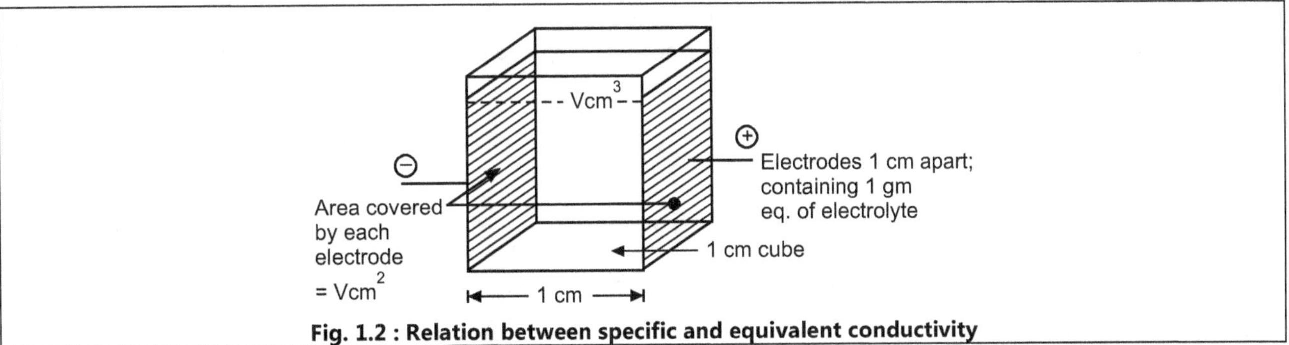

Fig. 1.2 : Relation between specific and equivalent conductivity

1.6 MOLAR CONDUCTANCE

(A) Electrical conductance (G) of a solution: It is the reciprocal of its resistance.

$$G = \frac{1}{R}$$

Unit of G: Ω^{-1} as R is in ohm (Ω), in SI system it is designated as siemens, S (1 S = 1 Ω^{-1}).

$$1\,S = 1\,\Omega^{-1} = \frac{Ampere}{Volt} = \frac{A}{V} \qquad \left(A = \frac{Coulomb}{Time} \text{ i.e. } \frac{C}{s}\right)$$

$\therefore$ $$\boxed{1\,S = CV^{-1}s^{-1}}$$

The electrical conductance of a material is inversely proportional to its length and directly proportional to its cross-sectional area.

$$G = \frac{1}{R}, \quad \text{but } R \propto \frac{l}{a} \quad \therefore \ G \propto \frac{a}{l}$$

$\therefore$ $$G = k \cdot \frac{a}{l}$$

where, k is constant called as conductivity of the conductor.

$\therefore$ $$k = G \cdot \frac{l}{a} = \frac{1}{R} \cdot \frac{l}{a} \qquad \qquad \dots (1.5)$$

When $l = 1$ m, a = 1 m^2, then k = G.

Conductivity (k) (Definition): The conductance of a unit cube of material is called as conductivity.

SI unit of conductivity is Sm^{-1}, more common units are $\Omega^{-1}\,cm^{-1}$ or S cm^{-1}.

From equations (1.2) and (1.5),

$$k = \frac{1}{\rho}$$

The conductivity is the inverse of resistivity.

(B) Molar conductivity of an electrolyte: The conductivity of a solution depends on the number of ions present in it. The solution of higher concentration show higher conductivity due to presence of more ions than the solution of lower concentration.

To compare the conductivities of different solutions, George Kohlrausch introduced the concept of molar conductivity.

Molar conductivity ($\wedge$): Molar conductivity is the electrolytic conductivity k divided by the molar concentration C of the dissolved electrolyte.

$\therefore$ $$\wedge = \frac{k}{C}$$

SI unit of molar conductivity is

$$\frac{Sm^{-1}}{mol\ m^{-3}} = Sm^2\,mol^{-1}$$

Common units are $\Omega^{-1}\,cm^2\,mol^{-1}$ or S cm^2 mol^{-1}. Molar conductivity can also be defined as the conductance exhibited by 1 molar solution of an electrolyte placed between two electrodes 1 cm apart and large enough to contain between them all the solution.

If C is the concentration of a solution in mol dm^{-3} i.e. $\dfrac{C}{1000\ mol\ cm^{-3}}$ then volume of the solution containing 1 mol of solute will be 1000/C cm^3.

k is the conductance of 1 cm^3 of the solution, the conductance of 1000/C cm^3 will be 1000 k/C which is molar conductivity.

$\therefore$ $\wedge = kV = \dfrac{1000\ k}{C}$, where V is the volume of the solution in cm^3 containing 1 mole of dissolved substance i.e. cm^3 mol^{-1} and C concentration in mol L^{-1}.

1.7 VARIATION OF SPECIFIC CONDUCTANCE WITH CONCENTRATION

- The conductivity of an electrolytic solution is directly proportional to the concentration as it depends on the number of ions present per unit volume of the solution.

- On dilution, the degree of dissociation increases, number of ions increases, but actually number of ions per unit volume decreases. Hence conductivity decreases.

- As strong electrolytes are completely ionized in the solution with increase in concentration, the number of ions per volume of the solution increases, hence the conductivity.

- In weak electrolytes due to partial dissociation, increase in number of ions per unit volume with concentration is comparatively small, therefore, conductivity does not increase rapidly as in strong electrolytes.

1.8 VARIATION OF MOLAR CONDUCTANCE WITH CONCENTRATION

- The molar conductivity is the conductance of all the ions produced by one mole of electrolyte. On dilution the total number of ions increases due to increase in the degree of dissociation. Hence the molar conductivity of both strong and weak electrolytes increases with dilution.

- The relation between conductivity and molar conductivity is given by $\wedge = \dfrac{k}{C} = k \cdot V$. On dilution although k decreases, increase in volume V is more than compensated by the decrease in value of k.

- On dilution, the molar conductivity of strong electrolytes increases upto the maximum limiting value $\wedge_o$ i.e. value at zero concentration or at ∞ dilution. The molar conductivity reaches the value $\wedge_o$ in just 0.001 or 0.0001 M solutions of strong electrolytes but for weak electrolytes at these concentrations it is less than $\wedge_o$.

1.9 VARIATION OF EQUIVALENT CONDUCTANCE WITH DILUTION

Variation of $\wedge$ with $\sqrt{C}$: Friedrich Kohlrausch gave the relation,

$\wedge = \wedge_o - a\sqrt{C}$ where a is the constant, $\wedge_o$ is the molar conductivity at ∞ dilution. The graph of $\wedge$ of strong and weak electrolytes versus square root of the concentration is as shown in Fig. 1.3.

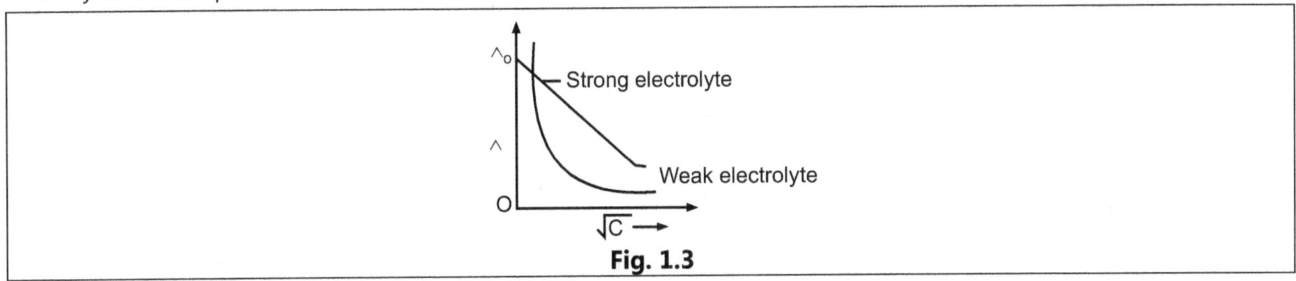

Fig. 1.3

- The molar conductivity of strong electrolytes show linear variation with $\sqrt{C}$. Weak electrolytes behave differently as shown in Fig. 1.3. The ∞ dilution means the solution is so dilute that further dilution does not increase the molar conductivity. The ions are apart, hence do not interact with one another.

- The extrapolation of linear graph of strong electrolyte gives $\wedge_o$ i.e. molar conductivity at zero concentration. But this is not applicable for weak electrolytes as the curve does not approach the linearity but $\wedge_o$ can be calculated by using Kohlrausch law.

Kohlrausch law of independent migration of ions : It states that at infinite (∞) dilution, each ion migrates independently of its co-ion and makes its own contribution to the total molar conductivity of an electrolyte irrespective of the nature of the other ion with which it is associated.

As both cation and anion of an electrolyte contribute to $\wedge_o$ i.e. molar conductivity at infinite dilution, $\wedge_o$ is given by

$\wedge_o = \lambda_+^o + \lambda_-^o$ where λ_+^o, λ_-^o are the molar conductivities of cation and anion at infinite dilution respectively.

For example: KI and NaI have $\wedge_o$ values 150.3, 126.9 respectively. The difference between $\wedge_o$ values is 23.4 which is equal to the difference between λ_o values of K^+ and Na^+ ions. This proves the validity of the law.

$$\wedge_o \text{ (KI)} - \wedge_o \text{ (NaI)} = \left(\lambda^o_{K^+} + \lambda^o_{I^-} \right) - \left(\lambda^o_{Na^+} + \lambda^o_{I^-} \right)$$

$$= \lambda^o_{K^+} - \lambda^o_{Na^+}$$

Application of Kohlrausch law:

By using this law, the molar conductivity of an electrolyte at zero concentration can be calculated, which is particularly important in calculating $\wedge_o$ of weak electrolytes where extrapolation method is not useful.

For example, $\wedge_o$ of CH_3COOH can be calculated from $\wedge_o$ values of strong electrolytes HCl, NaCl, CH_3COONa.

$$\wedge_o(CH_3COOH) = [\wedge_o(CH_3COONa) + \wedge_o (HCl)] - \wedge_o(NaCl)$$

$\wedge_o$ of strong electrolytes can be calculated by extrapolation method.

Conductivity and degree of dissociation of weak electrolyte:

The degree of dissociation of weak electrolytes is related to its molar conductivity at concentration C by an equation

$$\alpha = \frac{\wedge}{\wedge_o}$$

where, $\wedge_o$ – Molar conductivity at zero concentration

       $\wedge$ – Molar conductivity at concentration C

The dissociation constant k of weak electrolyte is given by

$$k = \frac{\alpha^2 \cdot C}{1 - \alpha}$$

Substituting the value of α,

$$k = \frac{(\wedge/\wedge_o)^2 \cdot C}{(1 - \wedge/\wedge_o)}$$

$$= \frac{\wedge^2 \cdot C}{\wedge_o^2 \left(\dfrac{\wedge_o - \wedge}{\wedge_o} \right)}$$

$$= \frac{\wedge^2 \cdot C}{\wedge_o(\wedge_o - \wedge)}$$

$$\therefore \qquad k = \frac{\wedge^2 \cdot C}{\wedge_o(\wedge_o - \wedge)}$$

1.10 MEASUREMENT OF CONDUCTIVITY

The conductivity, molar conductivity of a solution can be determined by measuring the resistance of a solution by Wheatstone bridge principle.

Conductivity cell: The cell is dipped in a solution whose resistance is to be measured as shown in Fig. 1.4.

The cell consists of a glass tube with two platinum plates coated with platinum black obtained by electrolysis of chloroplatinic acid.

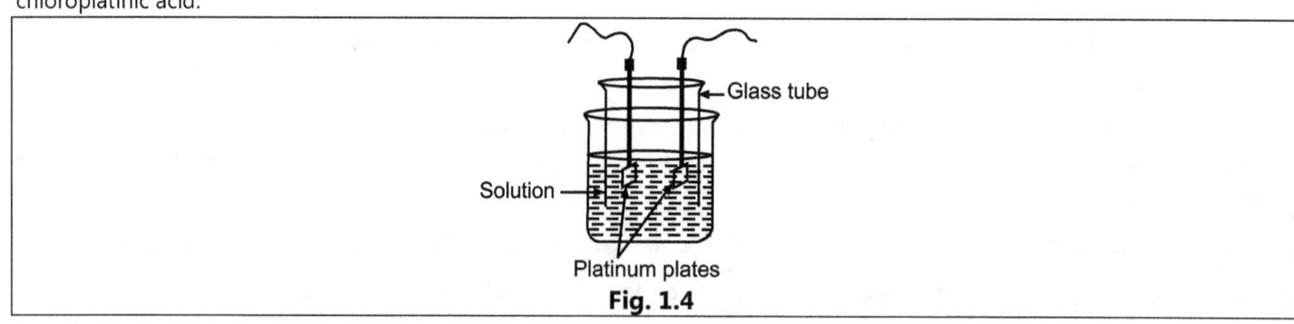

Fig. 1.4

FORMULAE

(1) $I = \dfrac{Q}{t}$ where, I = Current in amperes

(2) $V = IR$ Q = Charge in coulomb

(3) $R = \rho\dfrac{L}{A}$ V = Potential difference in volts

(4) Conductance $G = \dfrac{1}{R}$ R = Resistance in ohms

(5) Conductivity, $\sigma = \dfrac{1}{\rho}$ L = Length of a conductor

       A = Cross-sectional area of a conductor

       G = Conductance

SOLVED EXAMPLES

Example 1.1 :

A 6 m long copper wire has diameter 0.45 mm. If its resistance is 0.6 Ω, calculate its resistivity and conductivity.

Solution :

Given : $L = 6$ m, dia $= 0.45$ mm $= 0.45 \times 10^{-3}$ m, $\therefore r = 0.225 \times 10^{-3}$ m,

$$R = 0.6\ \Omega, \quad \text{Resistivity } \rho = ?, \quad \text{Conductivity } \sigma = ?$$

(i)   Resistivity, $\rho = \dfrac{R(A)}{L}$

$$= \dfrac{R\,(\pi r^2)}{L}$$

$$= \dfrac{0.6 \times [3.142 \times (0.225 \times 10^{-3})^2]}{6}$$

$$\boxed{\rho = 2 \times 10^{-8}\ \Omega m}$$

(ii)   Conductivity, $\sigma = \dfrac{1}{\rho}$

$$= \dfrac{1}{(2 \times 10^{-8})}$$

$$\boxed{\sigma = 5 \times 10^{7}\ S/m}$$

Example 1.2 :

A battery of emf 12 volt is connected across a resistance of 10 Ω. Calculate the current flowing through the resistance.

Solution :

Given : $V = 12$ V

     $R = 10\ \Omega$

     $I = ?$

We have, $V = IR$

$\therefore$ $I = \dfrac{V}{R} = \dfrac{12}{10}$

$$\boxed{I = 1.2\ A}$$

Example 1.3 :

A current of 0.6 A flows through a resistance of 25 Ω. Calculate voltage across it.

Solution :

Given :

$$I = 0.6 \text{ A}$$
$$R = 25 \ \Omega$$
$$V = ?$$
$$V = IR$$
$$V = 0.6 \times 25$$
$$\boxed{V = 15 \text{ V}}$$

Example 1.4 :

A current of 0.8 A flows through a resistance. If battery of emf 12 V is connected across it, calculate the resistance.

Solution :

Given :

$$I = 0.8 \text{ A}$$
$$R = ?$$
$$V = 12 \text{ V}$$
$$R = \frac{V}{I}$$

$\therefore$
$$R = \frac{12}{0.8}$$
$$\boxed{R = 15 \ \Omega}$$

Example 1.5 :

An electric heater draws a current of 4 A when connected across 220 V supply. What current will it draw when connected across 180 V supply ?

Solution :

Given :

$$I_1 = 4 \text{ A} \qquad\qquad I_2 = ?$$
$$V_1 = 220 \text{ V} \qquad\qquad V_2 = 180 \text{ V}$$
$$R = \frac{V_1}{I_1} \qquad\qquad \text{Now } R = \frac{V_2}{I_2}$$
$$= \frac{220}{4} \qquad\qquad \therefore I_2 = \frac{V_2}{R} = \frac{180}{55}$$
$$\boxed{R = 55 \ \Omega} \qquad\qquad \boxed{I_2 = 3.27 \text{ A}}$$

Example 1.6 :

The specific resistance of the material of a cable is 2.81×10^{-7} Ωm. If the resistance of the cable is 2.1 Ω and its radius is 0.8 mm, calculate the length of the cable.

Solution :

Given :

$$\rho = 2.81 \times 10^{-7} \ \Omega\text{m}$$
$$R = 2.1 \ \Omega$$
$$r = 0.8 \text{ mm} = 0.8 \times 10^{-3} \text{ m}$$
$$L = ?$$

We have,
$$R = \frac{\rho L}{A}$$

$\therefore$
$$L = \frac{R \ (A)}{\rho} = \frac{R \ (\pi r^2)}{\rho} = \frac{2.1 \times [3.142 \times (0.8 \times 10^{-3})^2]}{(2.81 \times 10^{-7})}$$
$$\boxed{L = 15.03 \text{ m}}$$

Example 1.7 :

Calculate the resistance of 60 m length of the wire having cross-sectional area of 0.02×10^{-6} m^2 and having resistivity 3.5×10^{-7} Ωm.

Solution :

Given :

$$L = 60 \text{ m}$$
$$A = 0.02 \times 10^{-6} \text{ m}^2$$
$$\rho = 3.5 \times 10^{-7} \text{ }\Omega\text{m}$$
$$R = ?$$

We have,

$$R = \frac{\rho L}{A} = \frac{(3.5 \times 10^{-7}) \times 60}{(0.02 \times 10^{-6})}$$

$$\boxed{R = 1050 \text{ }\Omega}$$

Example 1.8 :

The specific resistance of material of a wire is 3.2×10^{-7} Ωm. If the resistance of wire is 4 Ω and its length is 20 m, calculate the diameter of a wire.

Solution :

Given :

$$\rho = 3.2 \times 10^{-7} \text{ }\Omega\text{m}, \quad R = 4 \text{ }\Omega$$
$$L = 20 \text{ m}, \quad A = ?$$
$$\text{diameter} = ?$$

$$R = \frac{\rho L}{A}$$

$\therefore$

$$A = \frac{\rho L}{R} = \frac{3.2 \times 10^{-7} \times 20}{4}$$

$$A = 1.6 \times 10^{-6} \text{ m}^2$$

We have,

$$A = \frac{\pi d^2}{4}$$

$\therefore$

$$d^2 = \frac{4A}{\pi}$$

$\therefore$

$$d = \sqrt{\frac{4A}{\pi}}$$

$$\boxed{d = 1.4272 \times 10^{-3} \text{ m}}$$

Example 1.9 :

A resistance in the form of wire having length of 4 m and thickness 2 mm shows current of 500 mA for a potential difference of 12 volt. Calculate resistance in ohm and conductance in mho. Also calculate specific resistance of material of wire.

Solution :

Given : $L = 4$ m, Thickness diameter = 2 mm = 2×10^{-3} m. $\therefore$ r = 1×10^{-3} m

$I = 500$ mA = 500×10^{-3} A, $V = 12$ V, $R = ?$, $G = ?$, $\rho = ?$.

By Ohm's law, $\quad\quad V = IR$

$\therefore$

$$R = \frac{V}{I} = \frac{12}{(500 \times 10^{-3})}$$

$$\boxed{R = 24 \text{ }\Omega}$$

$$\text{Conductance, } G = \frac{1}{R} = \frac{1}{24}$$

$$\boxed{G = 0.0416 \text{ mho}}$$

We have,

$$R = \rho \frac{L}{A}$$

$\therefore$

$$\text{Specific resistance, } \rho = \frac{RA}{L} = \frac{R \times (\pi r^2)}{L} = \frac{24 \times 3.142 \times (1 \times 10^{-3})^2}{4}$$

$$\boxed{\rho = 1.885 \times 10^{-5} \text{ }\Omega\text{m}}$$

1.11 CELL CONSTANT

- The conductivity, k of an electrolytic solution is given by:

$$k = \frac{1}{R} \cdot \frac{l}{a}$$

The ratio $\frac{l}{a}$ is a constant quantity, called as cell constant 'b'.

Definition: The cell constant 'b' is the ratio of the distance between the electrodes to the area of cross-section of the electrode.

Cell constant, $\qquad\qquad\qquad\qquad b = \dfrac{l}{a}$

Unit: m^{-1} or cm^{-1}. $\qquad\qquad k = \dfrac{b}{R}$

Determination of cell constant: By using 1 M, 0.1 M or 0.01 M KCl solutions, cell constant is determined. The conductivity of KCl is known at different temperatures. By Wheatstone bridge principle, the resistance of KCl is determined.

AB is the uniform slide wire, R_x is the variable resistance placed in one arm of bridge as in Fig. 1.5. The conductivity cell dipped in KCl solution of unknown resistance is placed in another arm. D is the current detector (head phone or magic eye). F is the sliding contact that moves along AB.

A.C.: Alternating current source. The sliding contact F is moved across AB in such a way that current detector D shows no current. Suppose this is the point C.

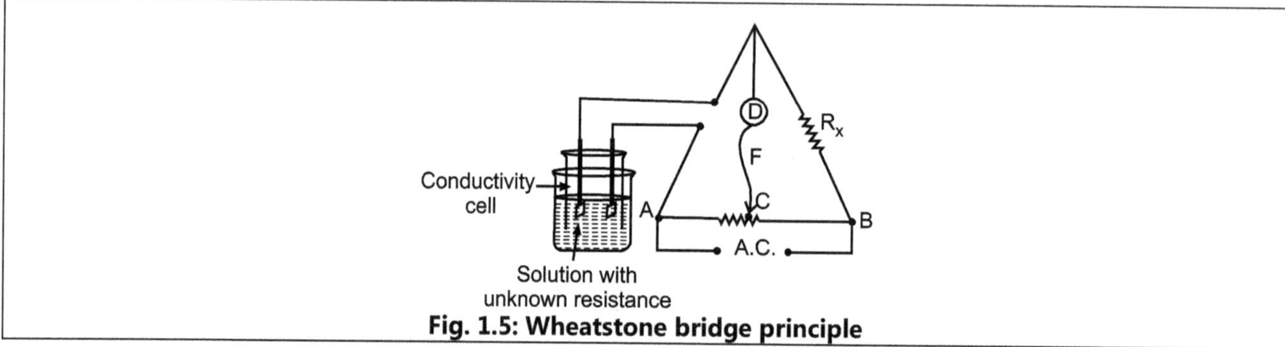

Fig. 1.5: Wheatstone bridge principle

According to Wheatstone bridge principle,

$$\frac{R_{(solution)}}{R(AC)} = \frac{R_x}{R(BC)}$$

But $\qquad\qquad\qquad R(AC) = l(AC), \ \ R(BC) = l(BC)$

$\therefore \qquad\qquad\qquad \dfrac{R_{(solution)}}{l(AC)} = \dfrac{R_x}{l(BC)}$

$\therefore \qquad\qquad\qquad R_{(solution)} = l(AC) \cdot \dfrac{R_x}{l(BC)}$

$\therefore \qquad\qquad\qquad R_{(solution)} = \dfrac{l(AC)}{l(BC)} \cdot R_x$

From the above formula, resistance of KCl solution can be calculated by measuring the lengths AC and BC, the value of R_x.

But $\qquad\qquad\qquad\qquad k = \dfrac{b}{R}$

$\therefore \qquad\qquad\qquad\qquad b = k \cdot R_{(solution)}$

The cell constant 'b' can be calculated by knowing the conductivity of solution, k.

Determination of conductivity of the given solution:

The given solution is placed in place of KCl solution and its resistance is measured by the Wheatstone bridge principle.

The conductivity, $\qquad\qquad\qquad k = \dfrac{b}{R}$

$\therefore$ Molar conductivity, $\wedge = \dfrac{k}{C}$, $\wedge$ can be calculated by knowing the concentration C.

1.12 CONCEPT OF pH AND POH

Hydrogen Ion Concentration : It can be readily seen that in any aqueous solution, the equilibrium

$$H_2O \rightleftharpoons H^+ + OH^-$$

will shift backward and forward according to the nature of solute but the equilibrium constant will remain constant at a fixed temperature, so that the ionic product of water, K_W, will also remain constant.

$$K_W = [H^+][OH^-] = 10^{-14}$$

In pure water, we have seen that $[H^+]$ equals $[OH^-]$ and this equality indicates that pure water is neutral. It is but quite obvious that in an acid solution, $[H^+]$ will be greater than $[OH^-]$ and in an alkaline solution, $[OH^-]$ will be greater than $[H^+]$.

Taking $[H^+] = [OH^-] = 10^{-7}$ mole dm^{-3} in neutral water or in neutral solution, we have

In acidic solution, $[H^+] > [OH^-]$ and hence

$$[H^+] > 10^{-7} \text{ and } [OH^-] < 10^{-7} \text{ mol/dm}^3$$

In alkaline solution, $[OH^-] > [H^+]$ and hence, $[OH^-] > 10^{-7}$ mol dm^{-3} and $[H^+] < 10^{-7}$ mol dm^{-3}. Thus, we can see that there is a very wide range from strongly acidic solution with H^+ ion concentration of about 1 mole per dm^3 to a strongly alkaline solution with H^+ ion concentration of about 10^{-14} mole per dm^3.

Definition of pH : *pH of a solution is defined as the negative logarithm to the base 10 of molar concentration of hydrogen ions or pH is the logarithm of the reciprocal of H^+ ion concentration as mol dm^{-3}.*

In symbol, pH of the solution, 'p' stands for *potenz*, meaning strength and pH indicates the strength of hydrogen ion concentration expressed in mol dm^{-3} in the solution. From the definition of pH, we have,

$$pH = -\log_{10}[H^+] = \log_{10}\frac{1}{[H^+]}$$

Above relation helps us to find out pH if $[H^+]$ is known and vice versa.

Definition of pOH : Just as pH is negative logarithm to the base 10 of $[H^+]$, *pOH is defined as the negative logarithm to the base 10 of $[OH^-]$, when the concentration is expressed in mol dm^{-3}.*

$$\therefore \quad pOH = -\log_{10}[OH^-] = \log_{10}\frac{1}{[OH^-]}$$

pH scale as a measure of Acidity and Alkalinity :

The method of expressing H^+ ion or OH^- ion concentration is very inconvenient to handle as it involves the negative power of 10. To decide the acidic or alkaline nature of a solution, it is necessary to express the hydrogen ion concentration $[H^+]$ on a convenient scale. Such a scale was suggested by Sorensen in 1909 and it is known as pH scale or Sorensen scale (Refer Fig. 5.10).

If $[H^+] = 1 \times 10^{-7}$ mol/dm^3, then

$$pH = -\log_{10}[1 \times 10^{-7}] = -(0-7) = +7$$

$$\therefore \quad pH = 7$$

Thus, for a neutral solution with $[H^+] = [OH^-] = 10^{-7}$ mol/dm^3, the pH is 7.

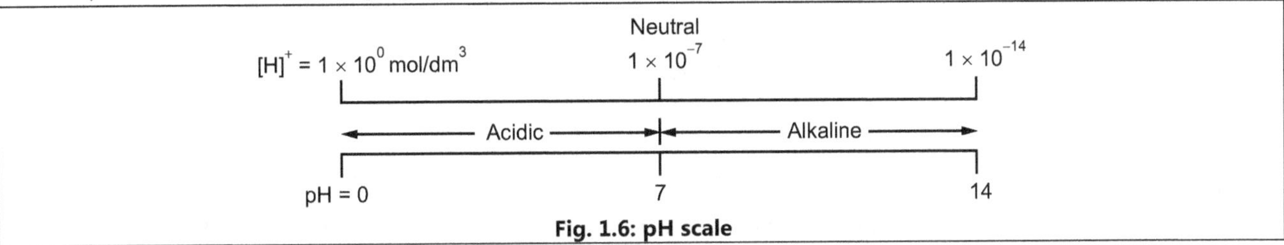

Fig. 1.6: pH scale

For an acid solution, $[H^+] > 10^{-7}$ mol/dm^3, the pH is less than 7. For alkaline solution with $[H^+] < 10^{-7}$ mol/dm^3, the pH is greater than 7. Variation in pH both for acidic and alkaline solutions can be shown diagrammatically as in Fig. 5.10. Thus from pH scale, nature of solution can be determined whether it is acidic, basic or neutral.

Derivation of pH + pOH = 14 :

In every aqueous solution,

At 298 K, $[H^+][OH^-] = 1 \times 10^{-14}$

Taking logs,

$$\therefore \quad \log[H^+] + \log[OH^-] = \log[1 \times 10^{-14}]$$

Changing signs, $-\log[H^+] + (-\log[OH^-]) = -(-14)$

$\therefore$ $pH + pOH = 14$

$$(\because pH = -\log_{10}[H^+] \text{ and } pOH = -\log_{10}[OH^-])$$

Hence, $pOH = 14 - pH$

The following Table 5.4 gives the relation between hydrogen ion concentration per dm^3 and pH and between hydroxyl ion concentration per dm^3 and pOH. Nature of the solution is also indicated.

Table 1.1

$[H^+]$	pH	$[OH^-]$	pOH	Ionic product	Nature of solution
1×10^0	0	1×10^{-14}	14	1×10^{-14}	Acidic
1×10^{-1}	1	1×10^{-13}	13	1×10^{-14}	Acidic
1×10^{-3}	3	1×10^{-11}	11	1×10^{-14}	Acidic
1×10^{-6}	6	1×10^{-8}	8	1×10^{-14}	Acidic
1×10^{-7}	7	1×10^{-7}	7	1×10^{-14}	Neutral
1×10^{-9}	9	1×10^{-5}	5	1×10^{-14}	Alkaline
1×10^{-11}	11	1×10^{-3}	3	1×10^{-14}	Alkaline
1×10^{-13}	13	1×10^{-1}	1	1×10^{-14}	Alkaline
1×10^{-14}	14	1×10^0	0	1×10^{-14}	Alkaline

*$\log_{10}[H^+]$ or $\log_{10}[OH^-]$ is commonly written as $\log[H^+]$ or $\log[OH^-]$ respectively.

In calculating pH of an aqueous solution, H^+ ions contributed by the dissociation of water should be ignored since their concentration is negligible when compared with that of H^+ ions contributed by the acid. Same is true for pOH value.

SOLVED EXAMPLES

Example 1.10 :

Calculate pH of a solution which contains 4×10^{-4} moles of H^+ ions dm^{-3}.

Solution : $pH = -\log_{10}[H^+]$

On substitution, $pH = -\log_{10}[4 \times 10^{-4}]$

$$= -[0.6021 - 4] = 4 - 0.6021 = \boxed{3.3979}$$

Example 1.11 :

Calculate pOH and pH values of 0.0016 M KOH solution assuming complete ionisation.

Solution :

In 0.0016 M KOH solution, the concentration of OH^- ions is 1.6×10^{-3} mole dm^{-3}.

$$pOH = -\log_{10}[OH^-] = -\log_{10}[1.6 \times 10^{-3}]$$

$\therefore$ $pOH = -(0.2041 - 3) = 3 - 0.2041 = 2.7959$

But $pH + pOH = 14$

$\therefore$ $pH = 14 - 2.7959 = \boxed{11.2041}$

Example 1.12 :

Calculate pH of 0.001 M H_2SO_4 solution, assuming complete dissociation.

Solution :

H_2SO_4 is a dibasic acid. It ionises as,

$$H_2SO_4 \longrightarrow 2H^+ + SO_4^{--}$$

$$\begin{array}{ccc} 0.001 & 0.002 & 0.001 \\ \text{mole} & \text{mole} & \text{mole} \end{array}$$

$$pH = -\log_{10}[0.002]$$

$$= -\log_{10}[2 \times 10^{-3}] = -(0.3010 - 3) = \boxed{2.6990}$$

Example 1.13 :

Calculate pH of 0.1 M acetic acid which is 1% dissociated.

Solution :

$$[H^+] = \alpha \times C$$

Here, $\quad \alpha = 0.01$ and $C = 0.1$

$\therefore \quad [H^+] = 0.01 \times 0.1 = 0.001 \text{ mol/dm}^3$

$$pH = -\log_{10}[H^+]$$
$$= -\log_{10}[1 \times 10^{-3}] = -(0-3) = \boxed{3}$$

Example 1.14 :

Find the values of $[H^+]$, $[OH^-]$, pH and pOH of 0.001 N solution of a monobasic acid. Assume that the acid is completely dissociated. $K_W = 1 \times 10^{-14}$ at 298 K.

Solution :

(i)
$$[H^+] = \alpha \times C$$
$$= 1 \times 0.001$$
$$= \boxed{1 \times 10^{-3} \text{ mol dm}^{-3}}$$

(ii) $[OH^-]$ ion concentration

We have,
$$[H^+][OH^-] = 1 \times 10^{-14} \text{ at 298 K}$$
$\therefore \quad [1 \times 10^{-3}][OH^-] = 1 \times 10^{-14}$

$\therefore \quad [OH^-] = \dfrac{1 \times 10^{-14}}{1 \times 10^{-3}} = \boxed{1 \times 10^{-11} \text{ mol dm}^{-3}}$

(iii) pH of the solution

$$pH = -\log[H^+]$$
$$= -\log[1 \times 10^{-3}]$$
$$= -(0-3) = 3$$

$\therefore \quad pH = \boxed{3}$

(iv) pOH of the solution

We have,
$$pH + pOH = 14$$
$\therefore \quad 3 + pOH = 14$

$\therefore \quad pOH = 14 - 3 = \boxed{11}$

1.13 BUFFER SOLUTIONS (Nov. 18)

- For many purposes, we have to maintain an aqueous solution at constant and specified pH value. There are many biological systems (e.g. blood) whose pH must not be allowed to vary considerably. The pH of human blood in a normal person is approximately 7.4. An increase or decrease of as much as 0.4 is likely to be fatal. For proper productivity of crops, the soil should have the proper pH. We need a solution that resists any tendency to change its pH.

Definition : *A solution which has a definite pH and which resists the sudden change in its pH even when a small amount of strong acid or base is added to it is known as buffer solution, or buffer.*

A buffer solution is also known as regulator solution or solution of reserve acidity and alkalinity.

Types of Buffer Solutions : There are two types of buffers :

(1) Acidic buffer

(2) Basic buffer.

Preparation of Buffer Solution :

(1) Acidic Buffer : Its pH is less than 7. It is prepared by dissolving in water a weak acid and its salt of strong base. Thus, following pairs of compounds can be used for preparing acidic buffers. (a) CH_3COOH and CH_3COONa. (b) $HCOOH$ and $HCOONa$. (c) C_6H_5COOH and C_6H_5COONa. (d) $HCOOH$ and $HCOOK$.

(2) Basic Buffer : Its pH is more than 7. It is prepared by dissolving in water a weak base and its salt of strong acid. Such a pair of compounds is NH_4OH and NH_4Cl.

Preparation of a Buffer Solution of desired pH : Consider a buffer solution containing a weak acid HA and its salt with a strong base.

The weak acid HA ionises as

$$HA \rightleftharpoons H^+ + A^-$$

By the law of mass action, we have,

$$K_a = \frac{[H^+][A^-]}{[HA]}$$

$$[H^+] = \frac{K_a[HA]}{[A^-]} \qquad \ldots (1.6)$$

Here, K_a = dissociation constant of acid HA.

On addition of a salt with anion A^-, the ionisation of acid is suppressed and the concentration of the anion is practically equal to that of the added salt. Therefore in the above equation.

[HA] = concentration of weak acid and $[A^-]$ = concentration of salt

$$\therefore \quad [H^+] = \frac{K_a[acid]}{[salt]}$$

$$\therefore \quad \log_{10}[H^+] = \log_{10} K_a + \log_{10}\frac{[acid]}{[salt]}$$

$$\therefore \quad -\log_{10}[H^+] = -\log_{10} K_a - \log_{10}\frac{[acid]}{[salt]}$$

$$\therefore \quad pH = pK_a + \log_{10}\frac{[salt]}{[acid]} \qquad \ldots (1.7)$$

This equation is known as **Henderson equation**. A buffer solution of desired pH can be prepared by taking the definite ratio of salt to acid. It is now possible to prepare universal buffer mixture by making suitable mixtures of acids and salts.

Suppose we wish to prepare a buffer of pH 5 using acetic acid ($K_a = 1.75 \times 10^{-5}$) and sodium acetate.

Now, $\qquad K_a = 1.75 \times 10^{-5}$

$$\therefore \quad pK_a = -\log K_a = 4.757$$

According to Henderson equation, $\quad pH = pK_a + \log_{10}\frac{[salt]}{[acid]}$

$$\therefore \quad 5 = 4.757 + \log_{10}\frac{[salt]}{[acid]}$$

$$\therefore \quad \log_{10}\frac{[salt]}{[acid]} = 5 - 4.757 = 0.243$$

$$\therefore \quad antilog\left(\log_{10}\frac{[salt]}{[acid]}\right) = antilog\ 0.243$$

$$\therefore \quad \frac{[salt]}{[acid]} = 1.75\ i.e.\ 1.75 : 1$$

Therefore, to prepare a buffer solution of pH 5, sodium acetate and acetic acid are to be used in the above ratio in moles.

$$pOH = pK_b + \log_{10}\frac{[salt]}{[base]} \qquad \ldots (1.8)$$

where, $\qquad K_b$ = Dissociation constant of a base.

But $\qquad pH + pOH = 14$

$$\therefore \qquad pH = 14 - pOH$$

Mechanism of Buffer Action : The property of the buffer solution to resist change in its pH value on addition of small amount of acid or base is known as buffer action.

Acidic Buffer : Consider a solution of CH_3COOH and CH_3COONa. The salt, CH_3COONa is almost completely ionised giving large concentration of CH_3COO^- ions and Na^+ ions. The weak acid, CH_3COOH is slightly ionised.

$$CH_3COOH \rightleftharpoons CH_3COO^- + H^+$$

$$CH_3COONa \longrightarrow CH_3COO^- + Na^+$$

The ionisation of CH_3COOH is much reduced due to common CH_3COO^- ions (common ion effect).

1. **Addition of acid :** When small quantity of an acid (even strong acid as HCl) is added to this solution, H^+ ions from acid combine with CH_3COO^- ions producing practically unionised CH_3COOH.

$$H^+ + CH_3COO^- \longrightarrow CH_3COOH$$

Thus, H^+ ions are used up and pH of the solution is not practically changed. The addition of acid is neutralised by CH_3COO^- ions. Thus, this buffer has reserve basicity due to presence of CH_3COO^- ions. The resistance to change in pH of a buffer on addition of small amount of acid is known as *reserve basicity*.

2. **Addition of alkali :** When small quantity of an alkali (even strong as NaOH) is added to the solution, OH^- ions from alkali combine with H^+ ions of acetic acid dissociating more and more of undissociated molecules of CH_3COOH and forming H_2O.

$$OH^- + CH_3COOH \longrightarrow CH_3COO^- + H_2O$$

Thus, OH^- ions are used up and pH of the solution remains practically unchanged. The addition of alkali is neutralised by acetic acid. *This buffer solution has reserve acidity due to CH_3COOH.* The resistance to change in pH of a buffer on addition of small amount of base or alkali is known as *reserve acidity*.

Thus, the same buffer solution has reserve acidity as well as reserve alkalinity.

Basic Buffer : Consider a solution of NH_4OH and NH_4Cl. NH_4Cl is almost completely ionised giving large concentration of NH_4^+ ions and Cl^- ions. The weak base, NH_4OH is slightly ionised.

$$NH_4OH \rightleftharpoons NH_4^+ + OH^-$$

$$NH_4Cl \longrightarrow NH_4^+ + Cl^-$$

The ionisation of NH_4OH is much reduced due to common NH_4^+ ions (common ion effect).

1. **Addition of base :** If a base (even strong like NaOH) is added, then OH^- ions from a base combine with NH_4^+ ions producing practically unionised NH_4OH.

$$OH^- + NH_4^+ \longrightarrow NH_4OH$$

Thus, OH^- ions are used up and pH of the solution remains practically unchanged. The addition of base is neutralised by NH_4^+ ions in buffer. This buffer solution has reserve acidity due to NH_4^+ ions.

2. If an acid (even strong like HCl) is added, then H^+ ions from an acid combine with OH^- ions of NH_4OH dissociating more and more of the undissociated molecules.

$$H^+ + NH_4OH \longrightarrow NH_4^+ + H_2O$$

Thus, OH^- ions are used up and pH of the solution remains practically unchanged. The addition of an acid is neutralised by NH_4OH. This buffer solution has reserve basicity due to NH_4OH.

Properties of Buffer Solutions :

1. It maintains its pH even when diluted with water.
2. It retains its pH even when kept for a long time.
3. It does not change its pH even when a small amount of strong acid or strong base is added to it.
4. It always has a definite pH.
5. It has few H^+ or OH^- ions.

Applications of Buffer Solutions :

Buffer solutions have numerous applications in various fields, such as analytical, industrial, pharmaceutical and biological systems. Following are some of the examples :

(1) Citric acid when added to milk of magnesia, magnesium citrate is formed and it acts as a buffer to stabilize milk of magnesia.
(2) Penicillin preparations are stabilized by the addition of sodium citrate which acts as a buffer.
(3) The uptake of oxygen by blood and its transport and distribution from lungs to various parts of the body is done at the pH of about 7.4. The electrolytes present in the blood act as a buffer solution to maintain the desired value of pH of the blood. Any alternation in the pH of blood would lead to serious condition of a patient.

(4) In the laboratory, many chemical reactions can be achieved only at definite pH value of the solution, e.g. the precipitation of ZnS from $ZnCl_2$,

$$ZnCl_2 + H_2S \longrightarrow ZnS + 2HCl$$

is not very efficient, since the H^+ ions formed accumulate in the solution and make the reaction to shift towards left. However, if H_2S is passed through $ZnCl_2$ solution in the presence of sodium acetate, whole amount of zinc precipitates as sulphide.

$$ZnCl_2 + H_2S \longrightarrow ZnS + 2H^+ + 2Cl^-$$
$$2CH_3COO^- + 2H^+ \longrightarrow 2CH_3COOH$$

Here, CH_3COO^- ions react with H^+ ions to produce slightly dissociated CH_3COOH molecules and the solution now contains a buffer of acetic acid and sodium acetate. This prevents the back reaction of H^+ ions on ZnS.

(5) Sodium benzoate is added as a buffer to preserve jellies and jams.

PRACTICE QUESTIONS

1. Define resistance and state its S.I. unit.
2. Define resistivity or specific resistance of a material and state its S.I. unit.
3. Define conductance and state its S.I. unit.
4. Define conductivity or specific conductance and state its S.I. unit.
5. State Ohm's law and state its equation with usual symbol meaning.
6. Define the following terms : (a) Ohm's law (in reference to electrolytes), (b) Specific conductivity, (c) Equivalent conductivity.
7. How is electrolytic conduction of an electrolyte determined experimentally ?
8. What is molar conductance and molar conductivity of an electrolyte ?
9. Explain the variation of specific conductance with concentration.
10. Explain the variation of molar conductance with concentration.
11. Explain the variation of equivalent conductance with dilution.
12. Explain the measurement of conductivity.
13. What is cell constant ? How can it be determined ?
14. Define specific conductance and equivalent conductance of an electrolyte. What is the relationship between the two ?
15. Explain the concept of pH and pOH.
16. Define buffer solutions.
17. Explain preparation of buffer solution with its mechanism.
18. Give the properties of buffer solutions.
19. Explain the applications of buffer solutions.
20. Define cell constant and state the mathematical relation between cell constant and specific conductance.

UNSOLVED PROBLEMS

1. A 5 m long wire has diameter 0.4 mm. If its resistance is 10 Ω, calculate its resistivity and conductivity.
   (**Ans.** (i) $\rho = 2.5 \times 10^{-7}$ Ωm, (ii) $\sigma = 0.4 \times 10^7$ S/m)
2. A battery of emf 6 V is connected across a resistance of 12 Ω, calculate the current flowing through the resistance.
   (**Ans.** I = 0.5 A)
3. A current of 0.8 A flows through a resistance of 30 Ω. Calculate the voltage across it. (**Ans.** V = 24 V)
4. A current of 1.2 A flows through a resistance if a battery of emf 8 V is connected across it. Calculate the resistance.
   (**Ans.** R = 6.67 Ω)
5. An electric equipment draws a current of 1 A when connected across 150 V supply. What current will it draw when connected across 220 V supply ? (**Ans.** $I_2 = 1.47$ A)
6. The specific resistance of the material of wire is 3×10^{-7} Ωm. If the resistance of the wire is 4 Ω and radius of the wire is 0.6 mm, calculate the length of the wire. (**Ans.** L = 15.08 m)
7. Calculate the resistance of 25 m length of wire having cross-sectional area 0.12×10^{-7} m^2 and resistivity 2.8×10^{-7} Ωm.
   (**Ans.** R = 583.3 Ω)
8. The specific resistance of material of a wire is 2.4×10^{-7} Ωm. If the resistance of wire is 6 Ω and its length is 18 m, calculate the diameter of a wire. (**Ans.** diameter = 9.57×10^{-4} m)

METALS AND ALLOYS

(A) METALS

2.1 DEFINITION OF METALLURGY

- *"Metallurgy is a process of extraction of metals from their ores economically and profitably".* The extraction of metals cannot be carried out by any other method, because extraction depends upon the nature and properties of metal. Consequently, different types of ores like oxides, sulphides, carbonates, sulphates etc. have to be treated differently for the recovery of pure metals from them. Generally, an ore is subjected to the following operations :

(A) Crushing or processing the ore, (B) Concentration, (C) Reduction and (D) Refining.

Following flow chart shows the different processes used in the extraction of metal from the ore :

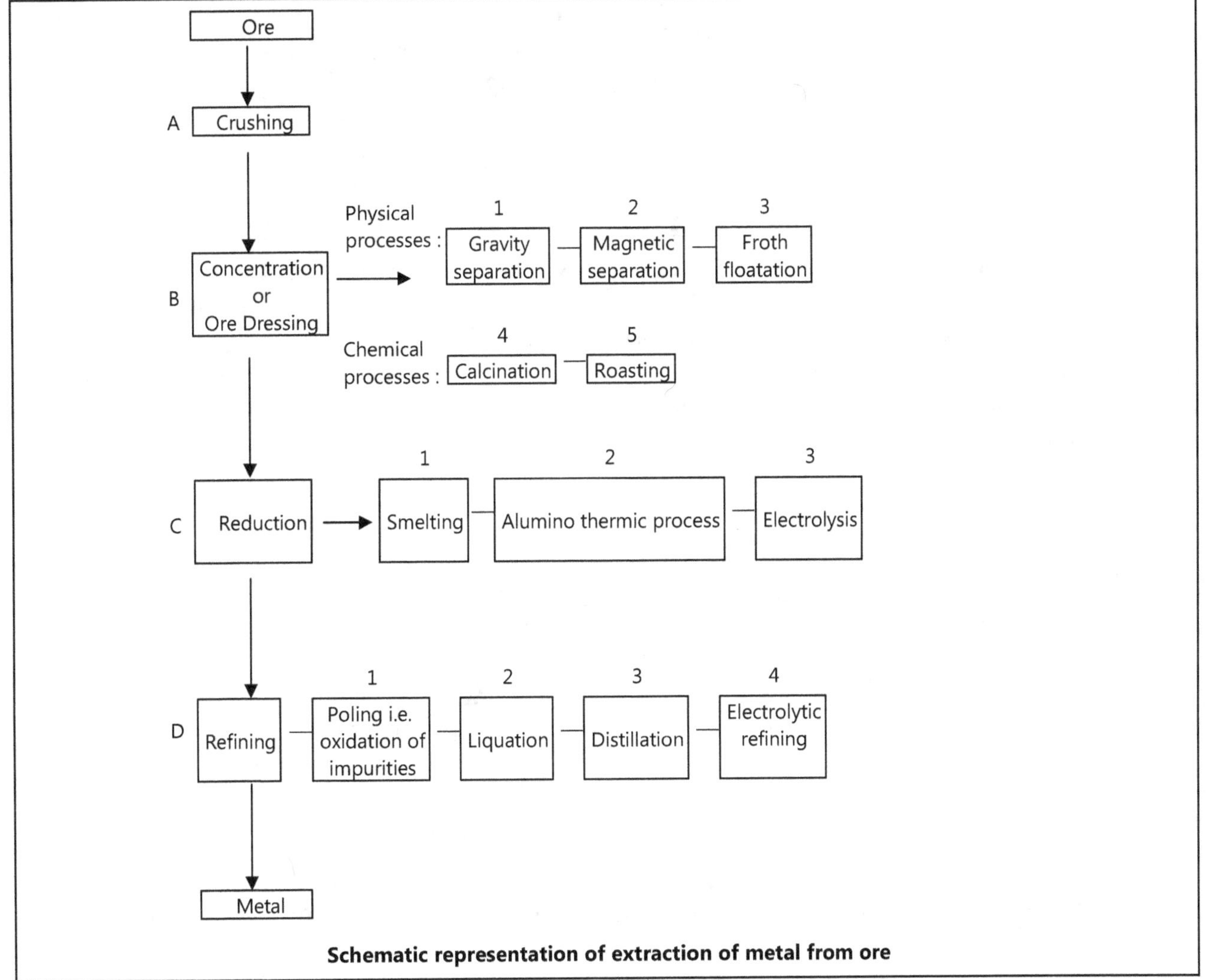

Schematic representation of extraction of metal from ore

(A) Crushing : The ore is obtained from mines in the form of big lumps (big stones). These lumps are unsuitable for further treatment. Hence, these are crushed to suitable size in big jaw crushers. These are then pulverised (i.e., converted into a fine powder) in ball mills or stamping mills.

(B) Concentration : The ore is generally associated with useless earthy or rocky impurities (like clay, sand etc.) called gangue or matrix. *"The process of removal of gangue or matrix from the ore is known as concentration of ore."* Thus, the percentage of metal in the concentration of ore can be brought about by the following methods depending upon the nature of ore.

(1) Gravity separation : In this process, the finely powdered ore is placed on a sloping platform and washed with a running stream of water. The lighter gangue particles are washed away, while the heavier ore particles sink at the depressions or at the bottom of sloping platform. (Refer Fig. 2.1) Generally, oxide ores like haematite (Fe_2O_3) and tinstone (SnO_2) are concentrated by gravity separation method.

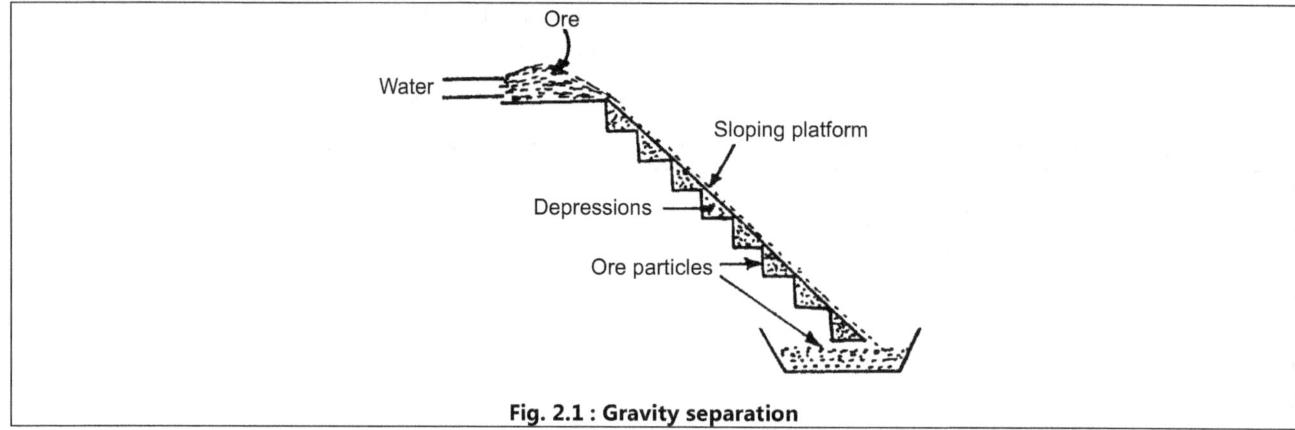

Fig. 2.1 : Gravity separation

(2) Electromagnetic separation : This process is used especially for separating magnetic impurities from non-magnetic ore particles. For example, tinstone (an ore of tin) in which tinstone (SnO_2) is non-magnetic, while the impurities like tungstates of iron and manganese are magnetic. The powdered ore (containing the magnetic impurities) is made to fall through the hopper on a non-magnetic belt of leather or rubber moving over the electromagnetic roller M. The magnetic impurities fall from the belt in a heap near the magnet, due to attraction; while the non-magnetic concentrated ore falls in a separate heap away from the magnet (Refer Fig. 2.2).

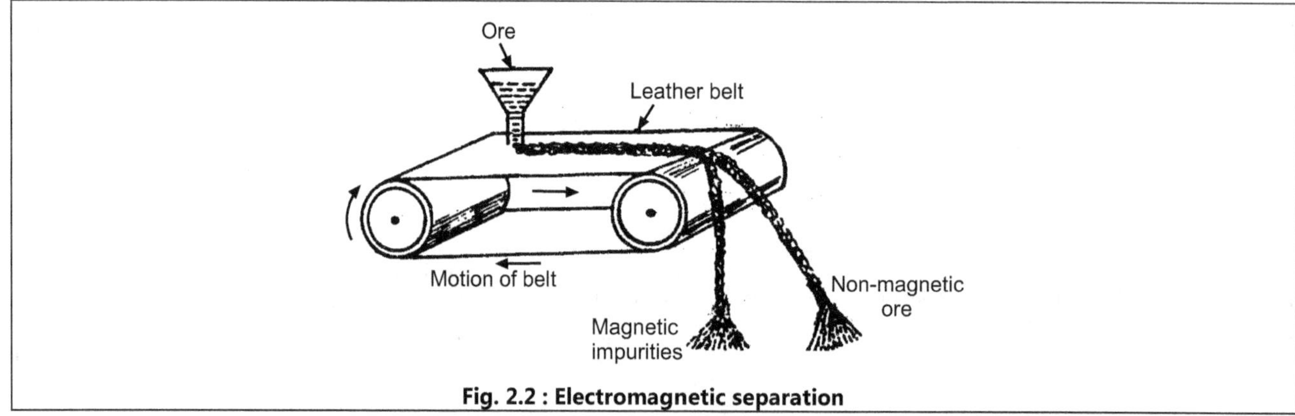

Fig. 2.2 : Electromagnetic separation

(3) Froth floatation process : This process is specially suitable for the concentration of sulphide ores like galena (PbS), nickel sulphide (NiS), zinc blende (ZnS), copper pyrites ($CuFeS_2$) etc. This process is based on the principle of different wetting characteristics of the ore and gangue particles with water and oil. The ore is preferentially wetted by oil and the gangue particles by water.

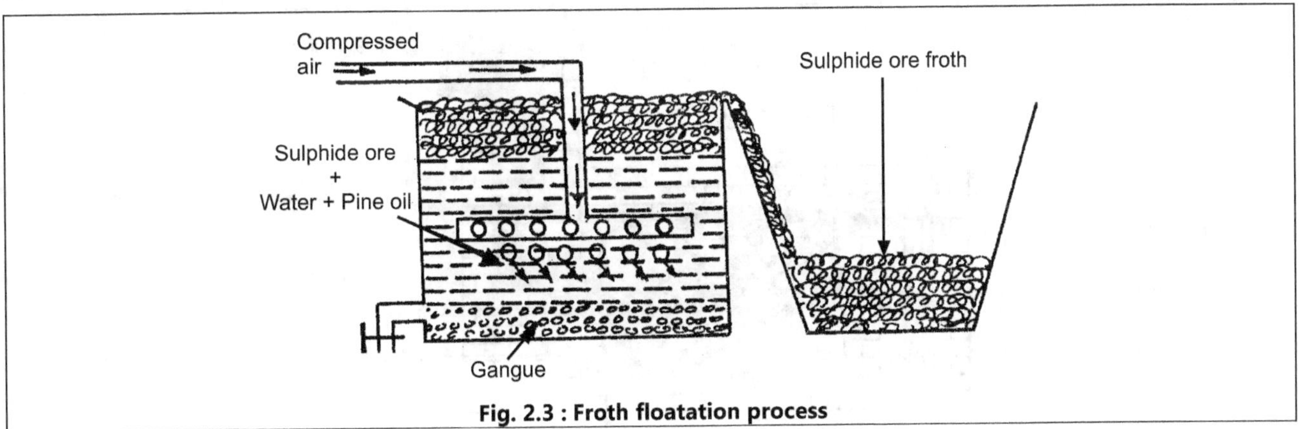

Fig. 2.3 : Froth floatation process

In this process, the powdered sulphide ore is mixed with water and pine oil. The whole mixture is then stirred vigorously by passing compressed air. The oil forms a froth with air bubbles. The sulphide ore particles get attached with the froth and floats on the surface, while the gangue or earthy impurities are wetted by water and sink to the bottom of the tank. The floating froth is then skimmed off into settling basins from where by filter press a concentrated ore is recovered.

(4) Calcination : *"Calcination is the process of heating the ore strongly in the absence of air to a temperature insufficient to melt it".*

Calcination is done in the hearth of a reverberatory furnace when the doors are kept closed (i.e. in the absence of air).

Generally, carbonate and hydroxide ores such as limestone ($CaCO_3$); malachite [$CuCO_3 \cdot Cu(OH)_2$] are concentrated by this method.

The main purposes of calcination are :

1. To convert carbonate and hydroxide ore into oxide.

$$CaCO_3 \longrightarrow CaO + CO_2 \uparrow$$

Lime stone

$$CuCO_3 \cdot Cu(OH)_2 \longrightarrow 2\ CuO + CO_2 \uparrow + H_2O \uparrow$$

Malachite

$$Fe_2O_3 \cdot 3\ H_2O \longrightarrow Fe_2O_3 + 3\ H_2O$$

Haematite

2. To remove the moisture.

3. To remove the volatile impurities.

4. To make the mass porous, so that it can be easily reduced to the metallic state.

(5) Roasting : *"Roasting is the process of heating the ore strongly in the presence of excess of air to a temperature insufficient to melt it".*

Roasting is done in the hearth of a reverberatory furnace when the doors are kept open for the free supply of air. This process is generally used in case of sulphide ores.

The main purposes of roasting are :

1. To convert sulphide into oxide and sulphate.

$$2\ ZnS + 3\ O_2 \longrightarrow 2\ ZnO + 2\ SO_2 \uparrow$$

$$ZnS + 2\ O_2 \longrightarrow ZnSO_4$$

2. To remove the moisture.

3. To remove volatile impurities like sulphur, arsenic, antimony and phosphorus in the form of their oxides.

4. To oxidise easily oxidisable substances.

It is difficult to reduce sulphide ore to a metal and hence it should be converted to its oxide before reduction.

Roasting is usually carried out in a special furnace known as a *reverberatory furnace* in which the ore is heated by flames of a fuel. These flames are reverberated, i.e. reflected from the arch-roof of the furnace. On one side of the furnace there is fire place and at the other side is an outlet (flue) for the waste gases. Air is admitted through the doors of the furnace.

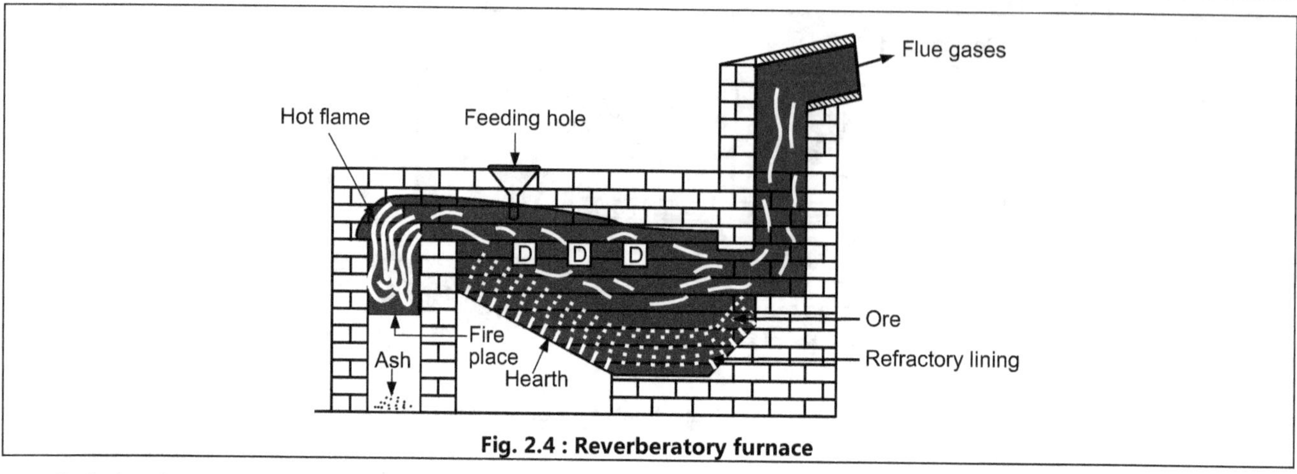

Fig. 2.4 : Reverberatory furnace

Similarly, when galena is roasted, a mixture of lead oxide and lead sulphate is formed.

$$2\,PbS + 3\,O_2 \longrightarrow 2\,PbO + 2\,SO_2 \uparrow$$

$$PbS + 2\,O_2 \longrightarrow PbSO_4$$

Distinction between Calcination and Roasting

Calcination	Roasting
1. It is the process of heating the ore strongly in the absence of air below its melting point.	1. It is the process of heating the ore strongly in the excess of air below its melting point.
2. Generally, this process is used to convert carbonate and hydroxide into their oxides.	2. Generally, this process is used to convert sulphide into oxide and sulphate.
3. The main object of calcination is to remove the moisture and volatile impurities from the ore.	3. The main object of roasting is to remove the moisture and oxidation of ore and the impurities like S, As etc.
4. In calcination, the mass becomes porous, so that it can easily be reduced to metallic state.	4. In roasting, the sulphide ore is chemically changed into a suitable form (oxide or sulphate) and can be reduced to metallic state.

(C) Reduction : This is the main part in the metallurgical operation (except the ores containing metals in native state). A reducing agent is used but it must be powerful and also cheap. Carbon in the form of coke is a satisfactory reducing agent. Other reducing agents like carbon monoxide or aluminium can also be used. Reduction is also carried out by an electric current.

Hence, there are three processes of reduction, namely :

(i) Smelting, (ii) Alumino thermic process and, (iii) Electrolysis.

(i) Smelting : Carbon reduces the oxide of a metal and the process is known as carbon reduction process. It is applicable in the cases of oxides of metals whose atomic weights are higher than that of manganese (55). Thus, zinc oxide, lead oxide, tin oxide, iron oxide etc., are reduced by carbon.

In this process, the calcined oxide ore is mixed with coke and flux (like CaO) and the mixture is strongly heated to high temperature. The reducing agent (coke) converts the ore into molten metal, while the flux removes the gangue in the form of fusible mass (slag). The process is also called *smelting* (to smelt means to melt). In some cases, the smelting is carried out in the reverberatory furnace at higher temperature while in many cases it is carried out in the blast furnace.

The hot blast of dry air is blown into the furnace just above the hearth through a number of pipes called *twyers*. In the well of the furnace there are two outlets known as *tap holes*. The upper tap hole is used to remove the slag and the lower one is used to remove the molten metal. Haematite ore is reduced by this method.

$$\underset{\text{Haematite}}{Fe_2O_3} + \underset{\text{Coke}}{3\,C} \longrightarrow \underset{\text{Iron}}{2\,Fe} + \underset{\text{Carbon monoxide}}{3\,CO \uparrow}$$

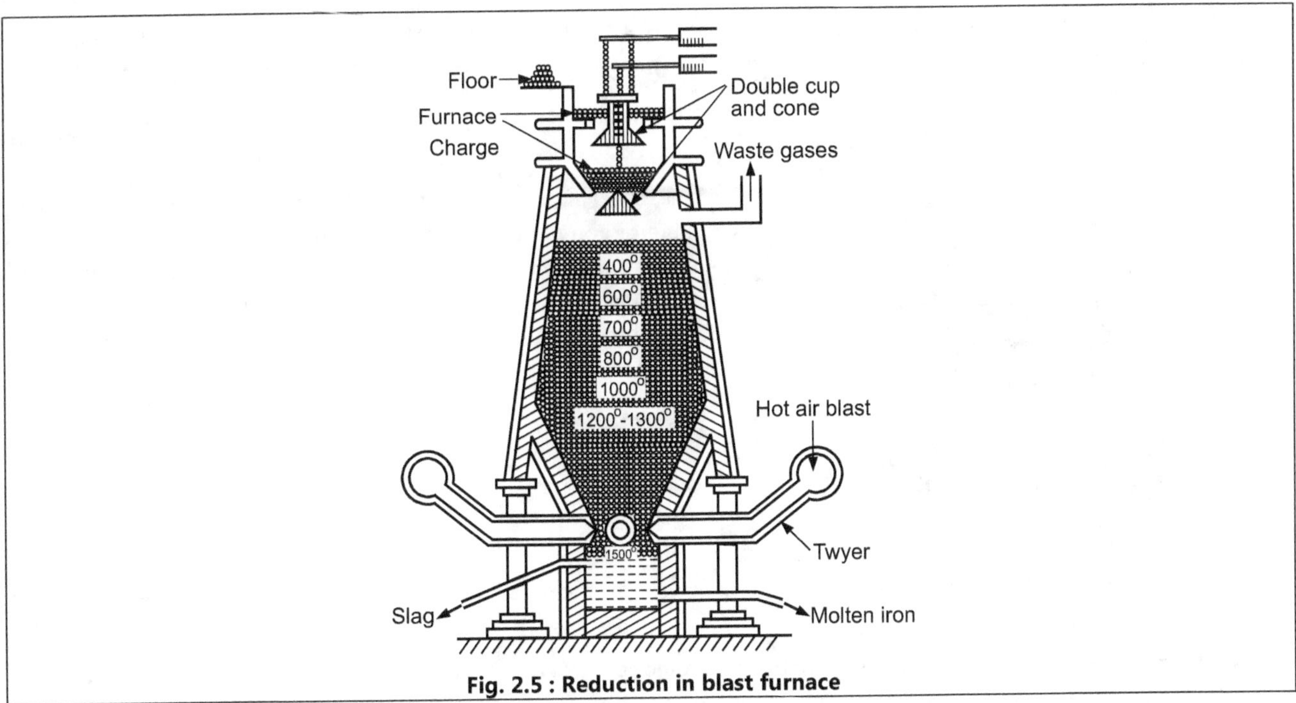

Fig. 2.5 : Reduction in blast furnace

Flux : In the above reduction of haematite, lime-stone is used as a flux. It decomposes at high temperature to give lime. This lime reacts with silica to form slag.

$$CaO + SiO_2 \longrightarrow CaSiO_3$$

　　　Flux　　Gangue　　Fusible slag

The various processes of concentration described above do not remove the gangue completely. To remove the gangue still present in the ore, certain oxides are added to the ore while melting it. Such *'a substance which is used to remove the gangue is called a 'flux'*. The flux combines with the gangue and forms an easily fusible compound which is called *slag*. The choice of a suitable flux depends upon the nature of the gangue. If the gangue is acidic in nature, CaO or MgO is used as a flux, which is basic in nature.

$$SiO_2 + CaO \longrightarrow CaSiO_3$$

　　(Acidic) Gangue　Flux　　　Slag

If the gangue is basic in nature, acidic flux is used i.e. iron oxide as a gangue is removed by the addition of silica as a flux.

$$FeO + SiO_2 \longrightarrow FeSiO_3$$

　　(Basic) Gangue　Flux　　　Slag

Hence, the terms gangue and flux in smelting are complementary to each other.

Slag : Even after concentration, the ore contains a considerable proportion of gangue which often contains oxides of very high melting point. These infusible oxides of metals and non-metals are converted into some fusible mass known as slag by the addition of suitable flux. Thus slag is the fusible chemical compound formed by combination of the added flux and the gangue (impurity) present in the ore. Therefore

$$Gangue + Flux \longrightarrow Slag \text{ (fusible compound)}$$
$$SiO_2 + CaO \longrightarrow CaSiO_3$$

　Silicious gangue　Flux　　　Slag

(ii) Aluminothermic process (or Thermite process) : If the oxides of a metal are very stable, aluminium is used as a reducing agent in place of carbon at high temperature. The oxides like Fe_2O_3 and Cr_2O_3 are very stable and can be reduced by aluminium powder at high temperature. A mixture of aluminium powder and metallic oxide (generally in the ratio 1 : 3 parts by wt.) is known as *thermite*. Aluminium at high temperature has a great affinity for oxygen. The reaction is exothermic, hence it liberates large amount of heat. Hence, aluminium can reduce metallic oxide to metal as shown by the following reactions :

(1) $Fe_2O_3 + 2\,Al \longrightarrow Al_2O_3 + 2\,Fe + Heat$

(2) $Cr_2O_3 + 2\,Al \longrightarrow Al_2O_3 + 2\,Cr + Heat$

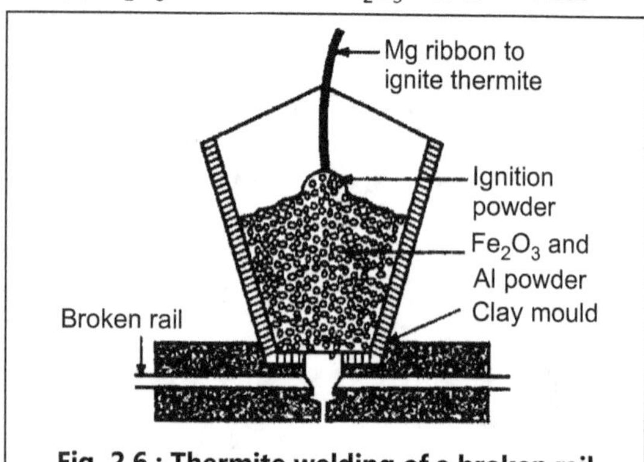

Fig. 2.6 : Thermite welding of a broken rail

[The ends of the broken rail are surrounded with a clay mould. A mixture of iron (III) oxide and aluminium powder (thermite) is ignited by magnesium ribbon in a funnel above. The molten iron produced by the reduction of iron (III) oxide runs into the mould. The molten iron melts the broken ends of the rail and produces a perfect union upon cooling.]

Metals obtained by this method are highly pure. Iron used in thermite welding, chromium and manganese needed for special steels are obtained by this method.

(iii) Electrolysis : This method is used in the case of oxide of very active metals like sodium, calcium, magnesium, aluminium, etc. which are not reducible by carbon. Hence, fused halides of such metals are electrolysed, when the metal is liberated at the cathode and halogen (Cl_2, Br_2, I_2, F_2) is liberated at the anode. Metallic sodium, potassium, calcium, magnesium etc, are obtained by this method. These metals cannot be obtained from their aqueous salt solutions by electrolysis as H^+ ion is discharged at cathode in preference to metallic ion. If more voltage is applied then the metal which may be obtained by electrolysis of fused sodium hydroxide or sodium chloride, aluminium by electrolysis of alumina in fused cryolite (Na_3AlF_6).

(D) Refining : After reduction it is necessary to remove small amounts of other elements that are usually present in that metal (except when the metal is obtained by electrolysis).

The process of purification of metal to get extra-pure metal is known as refining.

There are several ways of refining metal depending upon the nature of the metal to be purified. Some of the refining methods are

(i) Poling, (ii) Liquidation, (iii) Distillation, (iv) Electrolytic refining.

(i) Poling (i.e. oxidation of impurities) : This method consists of stirring the hot crude molten metal with green logs of wood. The wood gases (hydrocarbons like methane etc.) so produced reduce any metal oxide impurity present to the metallic form. Moreover, during stirring large quantities of air is absorbed by the molten metal and such absorbed air oxidises the easily oxidisable impurities. The oxidised impurities escape either as vapour or form 'scum' over molten metal. The scum so formed is removed by perforated ladle e.g. blister copper is refined by poling. The impurity is oxidised by absorbed air to As_2O_3, which escape away as vapour.

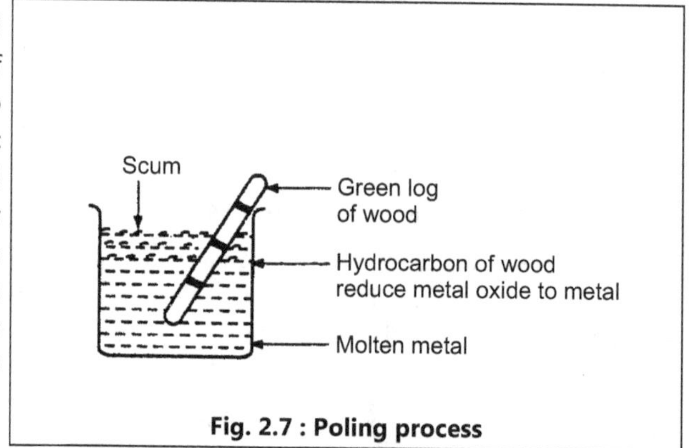

Fig. 2.7 : Poling process

(ii) Liquidation : This method is useful for metals having comparatively low melting points (like Pb, Sn etc.). The crude metal is heated on the sloping hearth of a furnace. The metal in the pure form melts first and flows down the slope leaving the infusible impurities behind called *"dross"*.

(iii) Distillation : The volatile metals (such as Hg, Zn, etc.) are purified by this method. When the crude metal is subjected to distillation, pure metal distils over leaving behind the high boiling point (non-volatile) impurities in the retort.

(iv) Electrolytic refining : *"The process of purification of metal to get extra pure metal with the help of electric current is known as electrolytic refining."* In this method, the anode consists of thick plates of impure metal and the cathode consists of thin plates of pure metal obtained by the previous operation or by chemical methods. The bath is an aqueous solution of a suitable salt of the metal which is to be refined. When an electric current with a definite voltage is passed through the bath, anode goes on dissolving and only the pure metal is deposited on the cathode which grows in size. Metals which are less electropositive than the one being refined settle below the anode and is known as *anode mud*.

Those metals which are more electropositive than the one being refined remain in solution as ions, as these ions accept electrons only at higher voltage. *Applied voltage is insufficient for these ions to be discharged at the cathode and thus the metal is refined.* This method is applicable to many metals such as copper, lead, nickel, zinc, silver etc.

2.2 METALLURGY OF IRON

2.2.1 Resources of Iron (Fe)

Chemistry of Iron:

Name	:	Iron (Prehistoric, from the Anglo-Saxon Iren or isern).
Symbol	:	Fe (From the Latin ferrum = iron).
At. No.	:	$Z = 26$.
At. Mass	:	55.85.
Ele. Conf.	:	2, 8, 14, 2; or $3d^6 4s^2$.
Valence	:	2 and 3.

Iron is the main transition element, belonging to 3d series, and present in the Gr. 8th (VIII B) and the 4th period of the periodic table.

A Word on Iron

Iron is most widely used popular metal from a pre-historical period. Its commercial and industrial importance is rather greater than any other metals. Number of advanced materials: such as synthetic rubber, plastic, etc. have been developed, but iron has maintained its supreme position in the industrial decorum.

Steel making is of immense importance throughout the world now and then. Today it is rather difficult to find out any side of modern living without involving iron or steels. Aeroplanes, boats, buildings, bridges, cookers, cars, railways, refrigerators, watches, utensils, etc. all contain iron and steels. Thus iron has occupied the most important place in our daily life. As a transitional element, iron is also the most important metal in plants and animals. Biologically it is important as an electron carrier in plants and animals (cytochromes and Ferrodoxins); as haemoglobin, the oxygen carrier in the blood or mammals, as myoglobin for oxygen storage, in nitrogenase (the enzyme in nitrogen fixing bacteria) and for iron scavenging and storage (ferritin and transferrin). It forms several unusual complexes, including ferrocene, etc.

The very importance of iron lies in its extra ordinary features such as:

(i)　It is the fourth most abundant element in the earth's crust. After O, Si and Al.

(ii)　It is very widely distributed in the earth's crust; and makes up 62,000 ppm or 6.2% by mass of the earth's crust; commonly associated with several hundred minerals.

(iii)　It is very easily isolated from its natural sources, simply by reducing the minerals with carbon.

(iv)　It has several desirable properties even in its impure form such as steels.

It is the cheapest, strongest, most magnetic and the most indispensable material from the constructional point of view.

Now-a-days iron and steel makings are regarded as the measures of technological developments of any country. Greatest stress has been laid by Government of India on the production of iron and steel, so that the country can industrialize quickly and extensively. At Jamshedpur, Bhilai, Durgapur, Rourkela, Bokaro and other place enormous quantities of cast iron and steel are being produced to meet the industrial and defence needs of India.

2.2.2 Important Ores of Iron

(Nov. 18)

Iron is wide spread in its occurrence. Its extent in earth's crust is 62,000 ppm or 6.2% by mass. It is 4th most abundant element on the earth, next to oxygen, silicon and aluminium. In meteorites, it occurs in free state while in nature, it is found in combined state such as oxides, sulphides, silicates etc. Its compounds occur in soil, rocks, minerals, green plants and in haemoglobin, the red colouring matter of blood.

The important minerals of iron are as follows:

(a) Oxides:

    (i)   **Magnetite (Black):** Fe_3O_4 or $FeO. Fe_2O_3$

       It is magnetic and black in colour. It is also called as *Ferrosoferric oxide*. It is exceedingly pure having high content (72.4%) of iron. It gives iron of high quality. It is found in large amounts in Canada, Sweden, USA and India. (In India it is found at A.P., Bihar, Chennai and Orissa). (The reserves estimated in India are 314 core tonnes of magnetite).

    (ii) **Haematite (Red): Ferric Oxide:** Fe_2O_3

       It has red colour and high percentage (~70%) of iron. It is found in Brazil, China, USA and India; (Bihar, Orissa, Maharashtra, (Chanda and Ratnagiri), Goa, Karnataka, M.P. and Tamilnadu. Estimated amount of Haematite is 960 crore tonnes in India).

    (iii) **Limonite (Brown Haematite):** $2Fe_2O_3\cdot 3H_2O$ or $Fe_2O_3\cdot xH_2O$.

       Hydrated ferric oxide. It is found in Belgium, Germany, France, USA and India. It contains ~ 60% iron.

(b) Carbonate: $FeCO_3$

Siderite or Spathic Iron ore: Ferrous Carbonate. It contains about 40-45% iron. It occurs in England and at West Bengal in India.

(c) Sulphides:

    (i)   **Iron Pyrites:** FeS_2 (Fool's Gold) (It is yellow metallic).

    (ii) **Copper Pyrites:** $CuFeS_2$

    (iii) **Arsenical Pyrites:** $FeAsS$.

Sulphur is an objectionable impurity as it causes iron to be hard and brittle and thus useless for technological purposes. The sulphides cannot be used for extraction of iron though they are plentiful.

The World production of iron ores was 970 million tonnes in 1988.

The largest sources are Russia 26%, China 17%, Brazil 15%, Australia 10%, the USA 6%, India 5%, and Canada 4%. From these the yield of pig iron was found to be 538 million tonnes in 1988.

Indian Seen:

The Indian iron ores contain 60 to 68% iron and are best in the world. The estimated total iron reserves of India are about 21000 million tonnes. This is regarded to be the largest deposit of high quality mineral source of iron in the whole world. The large deposits of iron ore are at Bengal, Bihar, Goa, Karnataka, Madhya Pradesh, Maharashtra and Tamilnadu.

Different Forms of Iron:

Iron is produced in three different commercial forms depending upon the amount of impurities present, especially carbon. They are:

Cast Iron	Steel	Wrought Iron
(Pig Iron)		(Pure form)
2.5 to 5% C	0.2 to 1.5% C	0.1 to 0.2% C

Iron is generally isolated as cast or pig iron by extracting it from oxide ore by simple reduction method using carbon as the reducing agent and lime as the flux in a blast furnace.

$$\text{Iron Oxide + Coke + Lime} \xrightarrow{\text{Blast Furnace}} \text{Cast Iron + Slag + Flue Gases}$$

In general, any alloy of iron containing more than 2% carbon is called as Pig Iron (where varying amounts of other impurities are also present). Pig iron is the direct metallic product of the blast furnace. It cannot be shaped into articles by forging or hammering. It is, therefore, melted and then cast into castings or the moulds of desired shape. It is then called Cast Iron. Thus, chemically, there is no much difference between cast iron and pig iron. Wrought iron also known as malleable iron is very pure form of iron containing not more than 0.5% carbon and useful for articles to be subjected to sudden stresses such as chains, wires, anchors, bolts, nails, etc.

The different forms of iron have different specific properties and hence they have different commercial and industrial applications.

2.2.3 Extraction Process

- The extraction of iron has played a very significant role in the development of modern civilization. Iron is produced in following steps:

Step I: Preliminary Treatment:

     (i)   Crushing and Grinding
     (ii)   Concentration
     (iii) Calcination or Roasting

Step II: Smelting (Reduction):   (Nov. 18)

Preliminary Treatment:

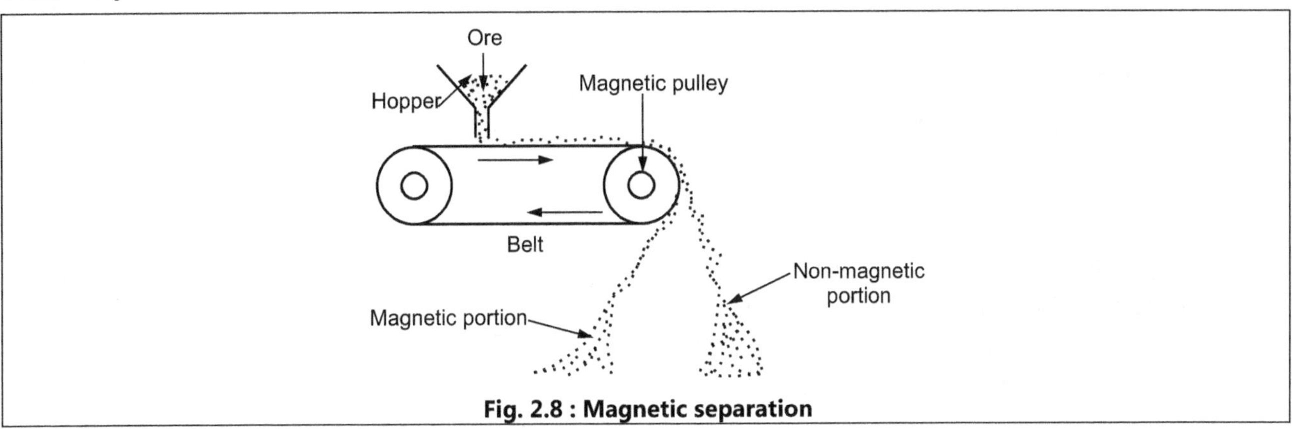

Fig. 2.8 : Magnetic separation

Concentration: Iron ore is first crushed and ground to suitable fineness, and then washed by using a powerful current of water (Gravity separation). The mass free from clay, sand and other earthy impurities are then dried and subjected to magnetic separation (Fig. 2.8). Thus the ore is concentrated to 90-95%.

Roasting:

The concentrated ore is roasted or calcined in a reverberatory furnace (Refer Fig. 2.9) at a low temperature, with a little coke; in free supply of air. By roasting following changes are brought about.

   (a)   Carbonate ore is converted to oxide:

$$FeCO_3 \longrightarrow FeO + CO_2 \uparrow$$

$$4\,FeO + O_2 \longrightarrow 2\,Fe_2O_3$$

   (b)   Ferrous oxide is oxidized to ferric oxide and thus its conversion to ferrous silicate (slag) is avoided.

$$4\,FeO + O_2 \longrightarrow 2\,Fe_2O_3$$

$$[FeO + SiO_2 \longrightarrow FeSiO_3 \text{ (Slag)}]$$

(c)   The mass is dried up by removing moisture.

$$3Fe_2O_3 \cdot 3H_2O \longrightarrow 2Fe_2O_3 + 3H_2O \uparrow$$

(d)   The lighter non-metallic impurities such as sulphur, arsenic, phosphorus etc. are all removed, as volatile substances.

$$S + O_2 \longrightarrow SO_2 \uparrow$$

$$P_4 + 5\,O_2 \longrightarrow P_4O_{10} \uparrow$$

$$4\,As + 5\,O_2 \longrightarrow 2\,As_2O_5 \uparrow$$

(e)   Ore becomes porous and hence easy to reduce.

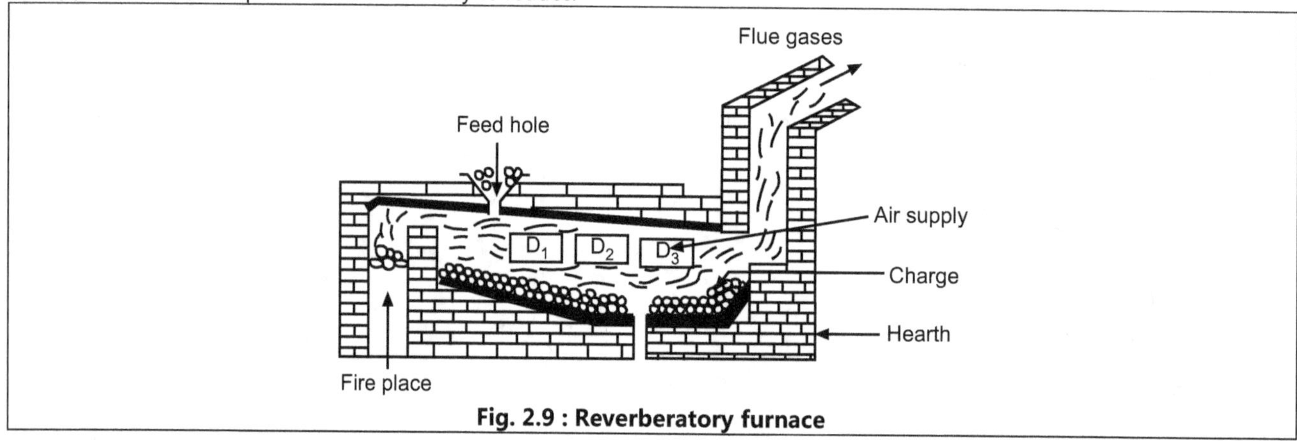

Fig. 2.9 : Reverberatory furnace

Smelting: Reduction:

The process of reduction of iron ore under molten condition is called smelting. Smelting is carried out in a blast furnace (designed in 1350 in Germany). It has two functions to perform.

(i)   To reduce the ore to metallic iron:

$$2Fe_2O_3 + 3C \longrightarrow 4Fe + 3CO_2 \uparrow$$

Or $$Fe_2O_3 + 3\,CO \longrightarrow 2Fe + 3CO_2 \uparrow$$

(ii)   To remove impurities in the form of slag:

$$CaCO_3 + SiO_2 \longrightarrow CaSiO_3 + CO_2 \uparrow$$

Or $$SiO_2 + CaO \longrightarrow CaSiO_3 \,(Slag)$$

Blast Furnace:

Blast furnace is a huge chimney - like tower constructed by steel plates. It is about 20 to 30 m in height and 4 to 8 m in width at the central part. It is lined with fireclay bricks from inside. The furnace is narrow near the mouth. (Refer Fig. 2.10) Such shape of the furnace facilitates wide spread of the charge inside the furnace and helps easy escape of the flue gases. This also makes proper heat flow. At the top, it is provided with double "bell" or "cup and cone" arrangement to permit charging of furnace without escape of waste gases. The flue gases are forced to escape only through the outlet, provided near the top of the furnace. The upper part of the furnace is called the throat, the middle part is known as the body and the bottom portion of the furnace is called the hearth. The furnace increases in diameter from the throat downwards until it attains a maximum diameter at the bosh. From the bosh downwards, furnace contracts more rapidly upto twyers and then it becomes nearly cylindrical. It retains this form even upto the bottom.

The lower part of the hearth called well of the furnace collects molten slag and molten iron. It is provided with two outlets. These are called tap holes. Molten slag is taken out from upper tap hole while molten iron is taken out from the lower tap hole. (Refer Fig. 2.10).

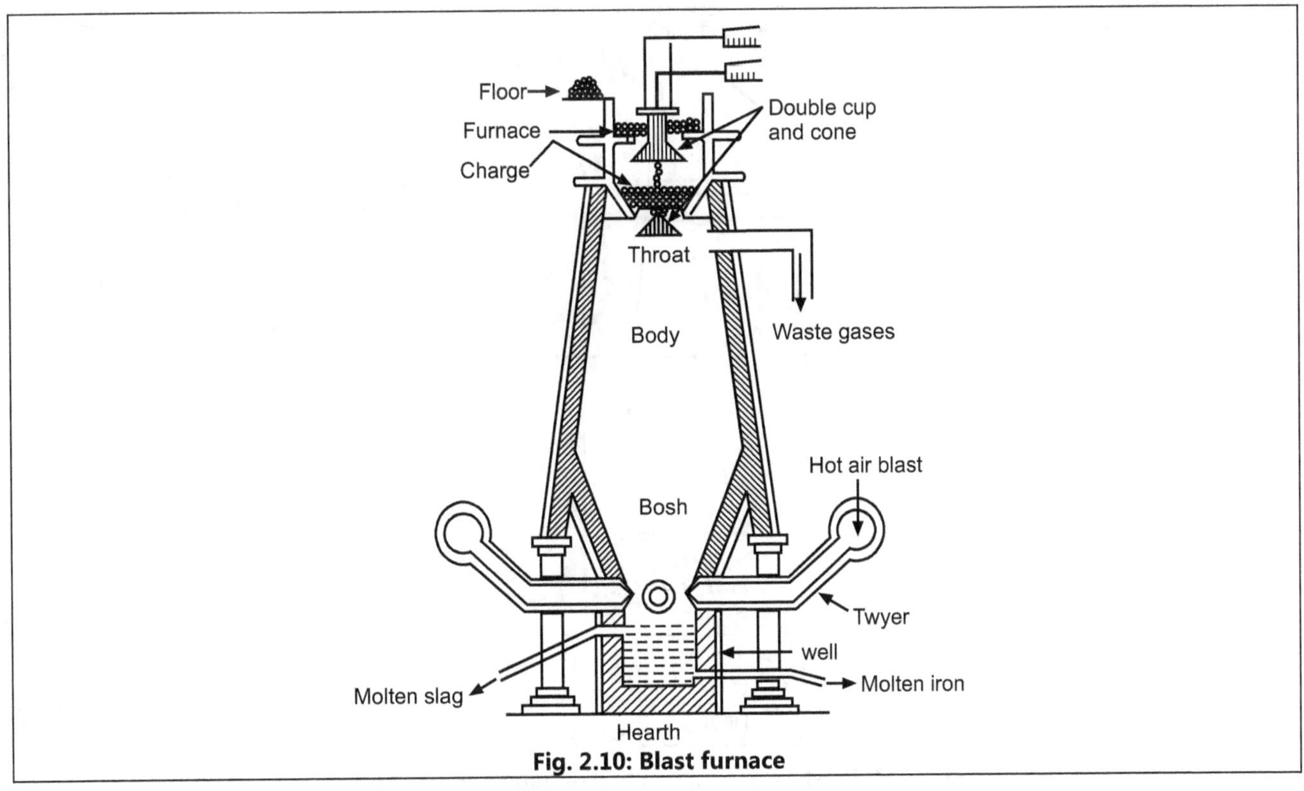

Fig. 2.10: Blast furnace

A little above the base, the furnace is provided with series of water jacketed pipes called bustles or twyers.

The hot blast of air, at temperature of about 800°C (1073 K) is blown into the furnace through the twyers at the rate of about 20,000 cu. m per minute and at a pressure of 1 kg/sq. cm. If the hot blast of air contains water vapour, a considerable absorption of heat (on account of the following endothermic reaction) may take place in the lower part of the furnace.

$$C + H_2O \longrightarrow CO + H_2 - 121 \text{ kJ mol}^{-1}$$

Hence air blast should not contain more than two per cent moisture. A nine to ten per saving of fuel is reported by drying the blast.

Working: The charge consisting (8 parts) of calcined ore, (4 parts) of desulphurised coke and (1 part) of limestone is carried to the top of the furnace and fed to it calculated amount periodically. Series of reactions occur in the furnace at different temperatures, 250°C (523 K) (near throat) to 1500°C (1773 K) (near hearth). (Refer Fig. 2.11).

Reactions in the Blast Furnace:

1. $3Fe_2O_3 + CO \rightarrow 2Fe_3O_4 + CO_2$
2. $Fe_3O_4 + CO \rightarrow 3FeO + CO_2$
3. $FeO + CO \rightarrow Fe + CO_2$
4. $CaCO_3 \rightarrow CaO + CO_2$
5. $Fe_2O_3 + 3C \rightarrow 2Fe + 3CO$
6. $CaO + SiO_2 \rightarrow CaSiO_3 \text{ (slag)}$
7. $SiO_2 + 2C \rightarrow Si + 2CO$
8. $Mn_2O_3 + 3C \rightarrow 2Mn + 3CO$
9. $2C + O_2 \rightarrow 2CO$

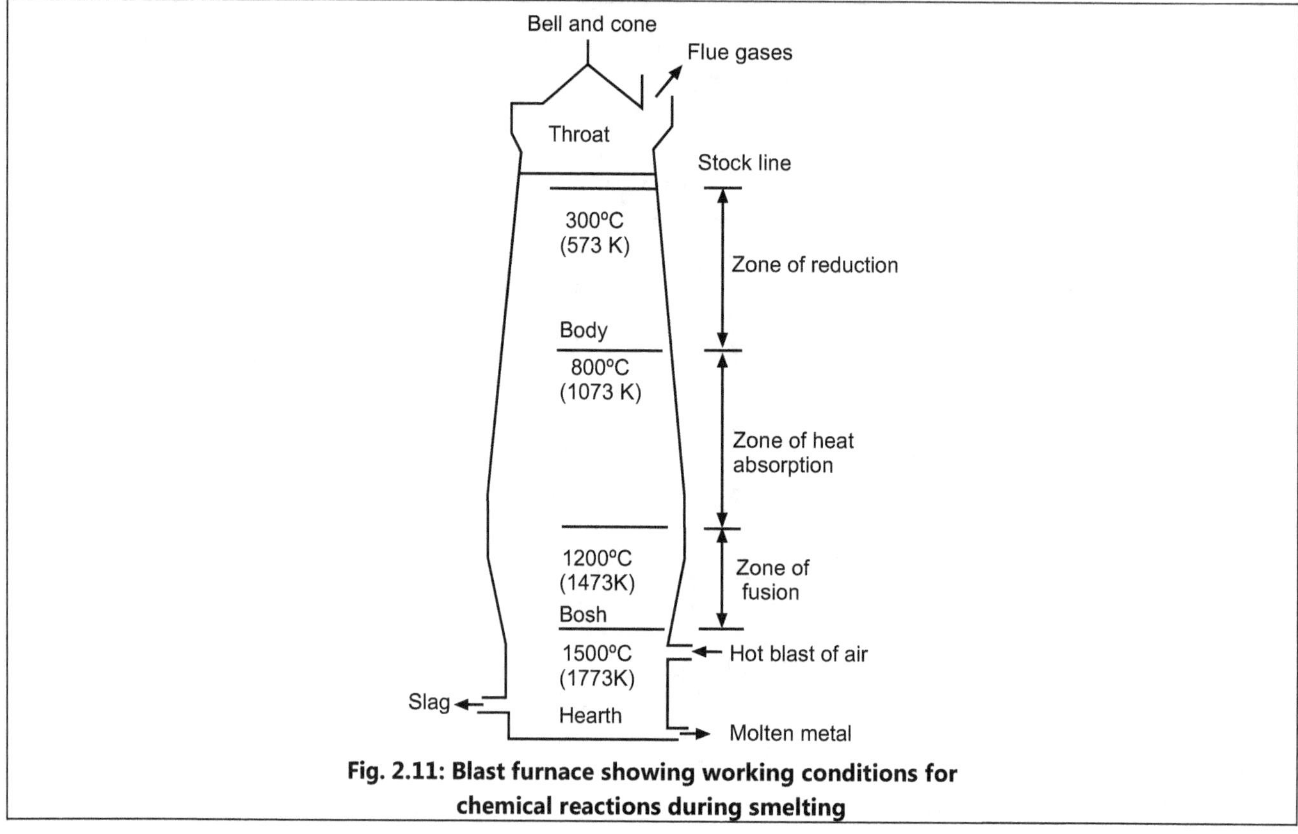

**Fig. 2.11: Blast furnace showing working conditions for
chemical reactions during smelting**

The hot blast of air passing through bustle oxidises coke to carbon dioxide. The reaction is exothermic and temperature rises to about 1500°C (1773 K) near the hearth.

$$C + O_2 \xrightarrow{\text{Oxidation}} CO_2 + 406 \text{ kJ mol}^{-1}$$

Carbon dioxide rises up in the furnace where it meets with the incoming coke. Here CO_2 reacts with coke and gets reduced to carbon monoxide.

$$CO_2 + C \longrightarrow 2CO - 163 \text{ kJ}$$

This reaction being endothermic, with ascending CO_2 the temperature goes on decreasing from 1500°C to 2500°C (1773 to 523 K) towards the neck of the furnace. Complicated reactions are found to occur at different temperatures as discussed.

1. **Zone of Reduction [(300° - 800°C) or (573 – 1073 K)]:**

   The main reaction that occurs near the top of the furnace is reduction of the iron oxide to metallic iron by carbon monoxide.

   $$Fe_2O_3 + 3CO \longrightarrow 2Fe + 3CO_2$$

   The process of reduction continues as the charge descends down. The reaction is not so simple but it is a series of reactions, taking place at different stages at different temperatures.

   $$3Fe_2O_3 + CO \xrightarrow{300^\circ C \text{ to } 500^\circ C} 2Fe_3O_4 + CO_2$$

   (At the top of the furnace)

   $$Fe_3O_4 + CO \xrightarrow{650^\circ C \text{ to } 700^\circ C} 3FeO + CO_2$$

   $$FeO + CO \xrightarrow{700^\circ C \text{ to } 800^\circ C} Fe + CO_2 \; \Delta H = -18.6 \text{ kJ mol}^{-1}$$

At the same, the limestone present in the charge is also decomposed to produce lime.

$$CaCO_3 \longrightarrow CaO + CO_2 \; \Delta H = 178 \text{ kJ mol}^{-1}$$

In short, iron oxide gets reduced in stages as follows:

$$Fe_2O_3 \longrightarrow Fe_3O_4 \longrightarrow FeO \longrightarrow Fe$$

The metal produced at first is spongy. Therefore, simultaneously with the process of reduction, a part of metallic iron reacts with carbon monoxide to form ferric oxide or ferroso-ferric oxide.

$$2Fe + 3CO \longrightarrow Fe_2O_3 + 3C$$

$$3Fe + 4CO \longrightarrow Fe_3O_4 + 4C$$

2. Zone of Heat Absorption [(800°C – 1200°C) or (1073 – 1473 K)]:

(i) In this region i.e. in body of furnace, unreacted iron oxide gets reduced to iron by red hot coke.

$$Fe_2O_3 + 3C \longrightarrow 2Fe + 3CO + 452 \text{ kJ mol}^{-1}$$

The hot spongy iron meets the ascending carbon monoxide.

(ii) Carbon monoxide gets disproportionate to CO_2 and carbon powder:

$$2CO \longrightarrow CO_2 + C \text{ (powder)}$$

This finely divided carbon powder gets mixed with iron in the finished product.

(iii) Lime obtained at the end of first zone combines with sand to produce slag.

$$CaO + SiO_2 \longrightarrow CaSiO_3 \quad \Delta H = 1450 \text{ kJ mol}^{-1}$$

Flux   Gangue Slag

(iv) Other impurities get reduced to the elementary substances and are mixed with finished iron:

(a) $MnO_2 + 2C \longrightarrow Mn + 2CO$

       Or $Mn_2O_3 + 3C \longrightarrow 2Mn + 3CO$

(b) $SiO_2 + 2C \longrightarrow Si + 2CO$

(c) $Ca_3(PO_4)_2 \longrightarrow 3CaO + P_2O_5$

(d) $CaO + SiO_2 \longrightarrow CaSiO_3 \text{ (slag)}$

(e) $P_2O_5 + 5C \longrightarrow 2P + 5CO$

Sulphur free from iron sulphide also mixes with iron. Thus mixing of C, P, S, Si, Mn etc. makes iron impure.

3. Zone of Fusion (Base of Furnace)[(1200°C – 1500°C) or (1474 – 1773 K)]:

Due to the presence of impurities, iron melts at 1200°C to 1300°C (1473 – 1573 K) though its M.P. is (1808 K) 1535°C. Molten iron then gets collected at the base called the well of the furnace. Slag also melts at the lower temperature (1200°C – 1250°C) (1473 – 1523 K) and being lighter than iron, it floats over molten iron and protects it from oxidation.

Slag is removed from the upper tap-hole while molten iron is removed from the lower tap hole, from time to time. The molten iron is then sent into sand moulds and cast into "pigs" or led to steel furnace and converted into different types of steels.

The extraction of iron by blast furnace is a continuous process where charge is fed from the top while metal is removed from the base periodically. The process runs day and night for years together. Everyday round about 1000 tonnes of cast iron and 500 tonnes of slag are removed at four different intervals.

(The modern blast furnace may be of 40×15 m in dimensions and can produce 10,000 tonnes of cast or pig iron per day.)

Products of the Blast Furnace

Cast iron (Pig Iron), slag and flue gases are the products of the blast furnace.

(a) Cast or Pig Iron:

The average composition of the cast or pig iron is as follows:

Iron	92 to 95%	Phosphorus	0.5 to 1%
Carbon	2.5 to 4.5%	Manganese	0.2 to 1%
Silicon	0.7 to 3%	Sulphur	0.1 to 0.3%

Cast iron is brittle and has low tensile strength. It melts at 1150°C – 1250°C (1423 – 1523 K). It can be forged, rolled and welded. It is therefore suitable only for casting not subject to sudden strain and shock. It is used for casting metal objects such as fire gates, pipes, railings, stoves, etc. Where low cast is more important than strength. When molten iron from the blast furnace is cooled suddenly, white crystalline cast iron is obtained. It is known as white cast iron and it contains iron carbide Fe_3C. On the other hand, if molten iron is slowly cooled, most of the carbon separates out as graphite, giving gray colour to the iron. It is softer variety and is called gray cast iron. Carbon present in the pig iron thus exists partly as free carbon in the form of graphite and partly in the combined state as iron carbide, Fe_3C.

(b) Slag:

Slag consists mostly calcium and aluminium silicates. It contains approximately 55% SiO_2, 30% CaO and 15% Al_2O_3. The slag is used as ballast (filler) for rail roads. It is also used in cement manufacture, as cement for road building, and as a fertilizer.

(c) Flue Gases:

The gases expelling out from blast furnace are called as flue gases. Flue gases have average composition 25% CO, 10% CO_2, 58 to 60% N_2, 4 to 5% H_2 and 1 to 2% hydrocarbons.

2.2.4 Engineering Applications of Pig Iron, Cast Iron, Wrought Iron or Malleable Iron

Applications of Pig Iron:

(1) Traditionally, pig iron was worked into wrought iron in finery forges, later puddling furnaces and more recently into steel. In these processes, pig iron is melted and a strong current of air is directed over it while it is stirred or agitated. This causes the dissolved impurities (such as silicon) to be thoroughly oxidized. An intermediate product of puddling is known as *refined pig iron, finers metal or refined iron.*

(2) Pig iron can also be used to produce gray iron. This is achieved by remelting pig iron, often along with substantial quantities of steel and scrap iron, removing undesirable contaminants, adding alloys and adjusting the carbon content. Some pig iron grades are suitable for producing ductile iron.

Modern uses :

(1) Until recently, pig iron was typically poured directly out of the bottom of the blast furnace through a trough into a ladle car for transfer to the steel mill in mostly liquid form; in this state, the pig iron was referred to as hot metal. The hot metal was then poured into a steel making vessel to produce steel, typically an electric arc furnace, induction furnace or basic oxygen furnace, where the excess carbon is burned off and the alloy composition is controlled. Earlier processes for this included the finery forge, the puddling furnace, the Bessemer process and the open hearth furnace.

(2) Modern steel mills and direct-reduction iron plants transfer the molten iron to a laddle for immediate use in the steel making furnaces or cast it into pigs on a pig-casting machine for reuse or resale. Modern pig casting machines produce stick pigs, which break into smaller 4-10 kg piglets at discharge.

Applications of Cast Iron :

(1) Cast irons are used in wide variety of applications owing to the properties like good fluidity, ease of casting, low shrinkage, excellent machinability, wear resistance and damping capacity.

(2) Applications :

- Car parts – cylinder heads, blocks and gearbox cases.
- Pipes, lids (manhole lids).
- Foundation for big machines (good damping property).
- Bridges, buildings.
- Cook wares – Excellent heat retention.

Applications of Wrought Iron :

(1) Wrought iron furniture has a long history, dating back to Roman times.

(2) It is also used to make home docor items such as baker's racks, wine racks, pot racks, etageres, table bases, desks, gates, beds, candle holders, curtain rods, bars and bar stools.

(3) The vast majority of wrought iron available today is from reclaimed materials. Old bridges and anchor chains dredged from harbors are major sources. The greater corrosion resistance of wrought iron is due to the siliccous impurities (naturally occurring in iron ore), namely ferric silicate.

(4) The use of wrought iron today is usually reserved for special applications, such as fine carpentry tools and historical restoration for objects of great importance.

(5) Wrought iron has been used for decades as a generic term across the gate and fencing industry, even though mild steel is used for manufacturing these 'wrought iron' gates. This is mainly because of the limited availability of true wrought iron. Steel can also be hot-dip galvanised to prevent corrosion, which cannot be done with wrought iron.

2.2.5 Metallurgy of Copper

(1) Metallurgy :

- The process of extracting a metal in its pure form from its ore is known as metallurgy.

- Copper is the second anciently known metal, the first being gold.

(2) Occurrence

- Copper occurs in native as well as in combined state. The principal ores are the sulphides and oxides.

1. **Sulphide ores :** (i) Copper pyrites : $CuFeS_2$, (ii) Copper glance : Cu_2S.

2. **Oxide ores :** Cuprite or ruby copper : Cu_2O.

3. **Carbonate ores :**

    (i) Malachite : $CuCO_3, Cu(OH)_2$, (ii) Azurite : $2CuCO_3, Cu(OH)_2$.

- In India, copper ores are found in Bihar (Singbhum), M.P., U.P., Mysore and Rajasthan (Khetrie, near Jaipur).

(3) Extraction

- Copper is extracted mostly from copper pyrites ($CuFeS_2$) by the dry process.

- The important steps involved in the process are as given below :

(i) Crushing : The copper pyrite ore is crushed in a big jaw crusher and then finally powdered in stamping mill.

(ii) Concentration :

    (A) Physical concentration by froth floatation.

    (B) Chemical concentration by roasting.

(A) Physical concentration :

- The finely powdered ore is concentrated by froth floatation process.

(B) Chemical concentration by roasting :

- The concentrated ore is heated strongly in the presence of an excess of air in a reverberatory furnace.

- During roasting, copper pyrites decompose to form cuprous and ferrous sulphides.

    $$2CuFeS_2 + O_2 \longrightarrow Cu_2S + 2FeS + SO_2 \uparrow$$

- A part of these sulphides get oxidised to corresponding oxides.

    $$2Cu_2S + 3O_2 \longrightarrow 2Cu_2O + SO_2 \uparrow$$
    $$2FeS + 3O_2 \longrightarrow 2FeO + SO_2 \uparrow$$

- Simultaneously, moisture is eliminated and impurities like sulphur, arsenic and antimony are removed in the form of their volatile oxides.

    $$S + O_2 \rightarrow SO_2 \uparrow$$
    $$4As + 3O_2 \rightarrow 2As_2O_3 \uparrow$$
    $$4Sb + 3O_2 \rightarrow 2Sb_2O_3 \uparrow$$

- Thus, the roasted copper ore contains a mixture of sulphides and oxides of copper and iron.

(iii) Reduction by Smelting :

- The roasted ore is mixed with coke and sand and heated in the presence of excess of air in a water jacketed blast furnace.

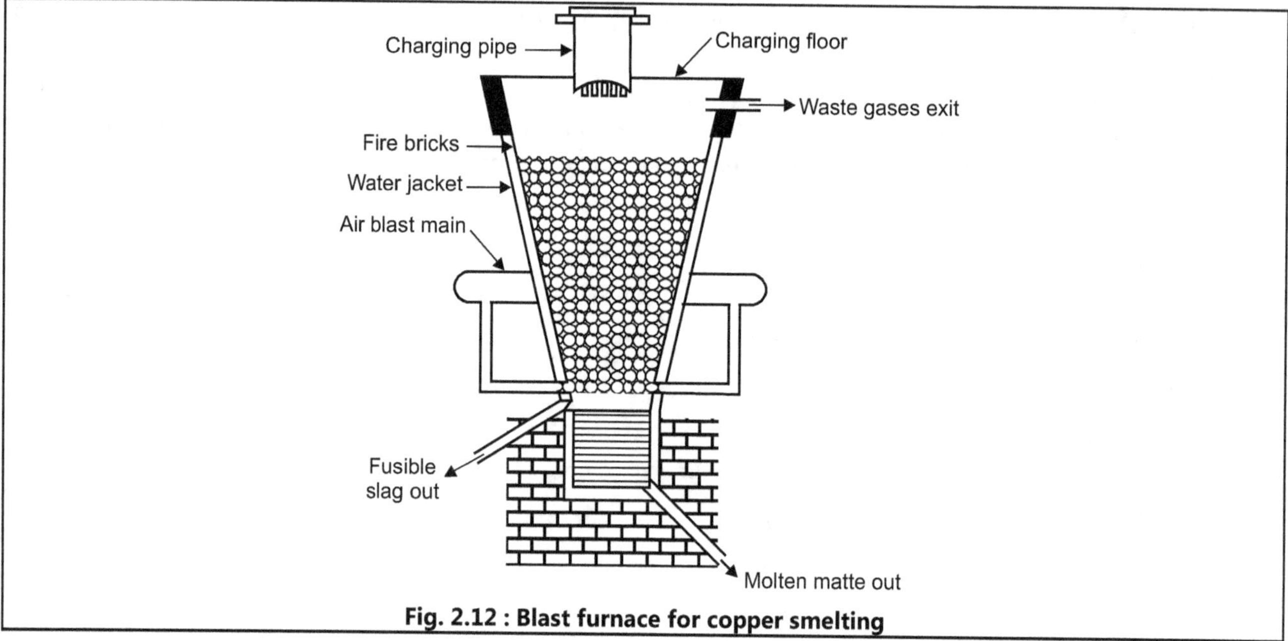

Fig. 2.12 : Blast furnace for copper smelting

- The modern blast furnace is about 15 to 20 feet high and is made of steel plates lined inside with fire bricks. It is water jacketed throughout and is provided with a waste gas outlet at the top.

- The roasted ore mixed with coke and sand is placed on the charging floor and is fed into the furnace through a charging pipe. (Refer Fig. 2.12).

- The oxidation of ferrous sulphide which started during roasting proceeds further and form ferrous oxide.

$$2FeS + 3O_2 \longrightarrow 2FeO + 2SO_2$$

- Ferrous oxide so formed, combines with sand to form fusible slag.

$$FeO + SiO_2 \longrightarrow FeSiO_3$$
$$\text{(slag)}$$

- Cuprous oxide formed during roasting combines with FeS and is changed back into its sulphide (because iron has great affinity for oxygen than copper).

$$Cu_2O + FeS \longrightarrow FeO + Cu_2S$$

- Thus, most of the iron is converted into the oxide, which is removed as slag. Slag being lighter, rises to the surface and removed from the slag hole.

- While the molten mass containing mostly cuprous sulphide with a little ferrous sulphide (remaining unchanged) called "matte" is taken out from the exit at the bottom. Thus, the matte is a mixture of molten Cu_2S + FeS.

- The smelting can also be carried out in a long reverberatory furnace. The reactions are similar to that in blast furnace.

- The matte (molten Cu_2S + FeS) so produced is then converted to blister copper by Bessemerisation.

(iv) Bessemerisation :

- The molten matte is now transferred to a Bessemer converter.

- It is a pear-shaped furnace made up of steel plates and lined with basic lining of lime or magnesia.

- It is mounted on trunnions and can be tilted in any position.

- The furnace is provided with pipes known as twyers through which sand and hot air is blown into it.
- The twyers are fitted in the sides (not in the base) and sufficiently high above the bottom so that the molten metal drops below the level of twyers and escapes the oxidising action of the air.

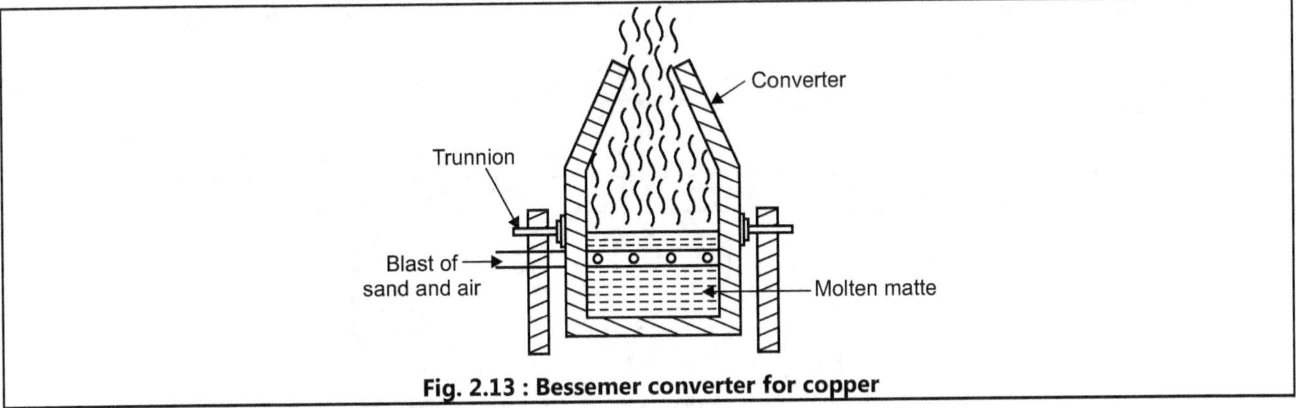

Fig. 2.13 : Bessemer converter for copper

- Following reactions take place in the Bessemer converter :

(a)   Conversion of FeS to slag.

$$2FeS + 3O_2 \rightarrow 2FeO + 2SO_2 \uparrow$$

$$FeO + SiO_2 \rightarrow FeSiO_3$$

$$\text{slag}$$

(b)   Partial oxidation of Cu_2S to Cu_2O.

   (A part of Cu_2S is oxidised to Cu_2O)

$$2Cu_2S + 3O_2 \rightarrow 2Cu_2O + 2SO_2 \uparrow$$

(c)   Reduction of Cu_2O by Cu_2S to metallic copper.

   (Rest of Cu_2S combines with Cu_2O to form blister copper)

$$2Cu_2O + Cu_2S \rightarrow 6Cu + SO_2 \uparrow$$

- Slag is poured off by tilting converter. This reaction is very vigorous and audible. The heat evolved is sufficient to keep the charge in the fused state.
- The molten metal is run off into sand moulds and allowed to stand. On cooling, dissolved SO_2 escapes out causing blisters on the surface of copper.
- Thus, the copper so produced is known as blister copper, which contains 96-98% Cu, rest being iron together with small amounts of Zn, Ni and As present originally in the ore.
- The blister copper is used as such for many purposes e.g. manufacturing of pipes, boilers etc. However, Cu required for conducting electricity must be perfectly pure. Hence blister copper is subjected to refining to get extrapure copper.

(v) Refining :

- The refining of blister copper is done by following two methods :

(1) Poling :

- In this process, blister copper is melted in a reverberatory furnace in the presence of air.
- The molten mass is kept stirred with green wood poles.
- Most of the impurities (S, As, Fe etc.) are removed either as volatile oxides (SO_2, As_2O_3) or as slag.
- Moreover, the cuprous oxide formed in the converter is reduced to metallic copper by reducing gases such as hydrocarbons of wood gases (like methane) produced from the green wood poles.
- By this refining, copper of 99.2-99.6% purity is obtained and it is known as tough pitch.

(2) Electrolytic refining :

- The tough pitch (containing 99.2-99.6% Cu) may be further refined to obtain 99.9% pure copper by electrolytic refining.
- Electrolytic refining is carried out in a large lead lined rectangular tank.
- The tough pitch (or impure copper) is cast into blocks which are made anode, which are suspended into the tank at intervals.
- Cathodes are thin plates of pure copper and each is suspended between two blocks of anode.
- The electrolyte consists of 15% $CuSO_4$ and 5-10% H_2SO_4 acid.
- The cathode and anode are connected either in multiple system or in series system.

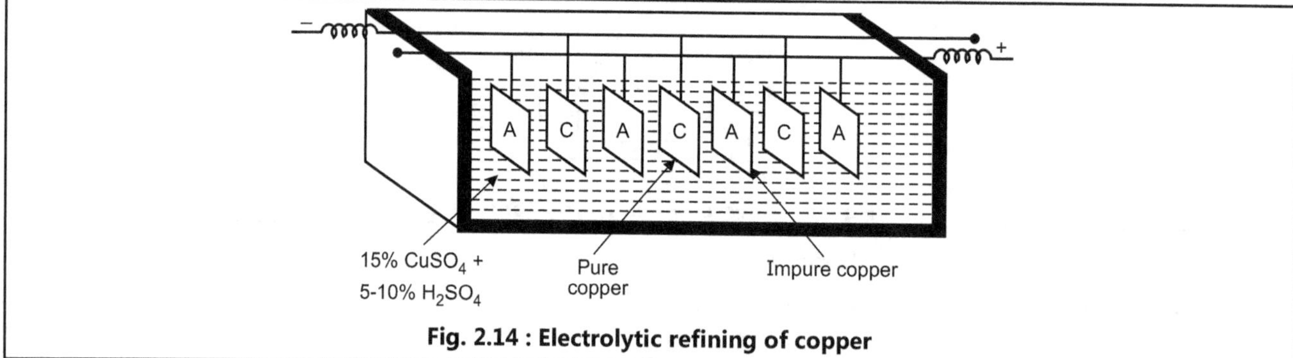

Fig. 2.14 : Electrolytic refining of copper

- On passing an electric current, copper from the crude anodes go into the solution and pure copper is deposited at the cathode.
- The impurities of more active metals (like Zn, Ni, Fe etc.) go into the solution as metallic ions.
- While impurities of less active metals (like Au, Ag, Pt etc.) are not ionised but crumbles down from the anodes and settle below anodes as 'anode mud'.

  From the anode mud, precious metals like Ag, Au, Pt are recovered. These costly metals pay the cost of the electrorefining process.
- At the applied voltage, Cu^{++} ions alone are discharged at the cathode (as copper is very low in the activity series of metals) and thus pure copper is deposited at the cathode.
- The cathodes grow in size and copper deposited can be removed from cathodes.
- The electrorefined copper is about 99.99% pure.

(4) Properties of Copper

Physical properties :

1. Copper is a heavy reddish brown metal of sp. gr. 8.93. It takes high polish easily.
2. It melts at 1083°C and boils at 2350°C.
3. It is malleable and ductile and can be drawn into wires of as small a diameter as 0.03 mm.
4. Copper is next to silver in conductivity of heat and electricity. However, the presence of other metals, even in traces, lowers the conductivity appreciably, e.g. the presence of 0.3% As lowers the conductivity of copper by 15%.
5. Molten copper absorbs SO_2 which causes blisters on cooling.
6. It casts very well and forms alloys (brass, bronze) easily.

Chemical properties :

1. **Action of air :** Copper is not attacked by dry air at ordinary temperatures. However, when heated to redness in air or oxygen, it first gives cupric oxide (CuO), which on further heating (above 1100°C) is converted to cuprous oxide (Cu_2O).

$$2Cu + O_2 \rightarrow 2CuO$$

$$2CuO + 2Cu \xrightarrow[1000°C]{above} 2Cu_2O$$

$$\text{or} \quad 4Cu + O_2 \xrightarrow[1000°C]{above} 2Cu_2O$$

In moist air and in presence of CO_2, it is coated with a green layer of basic carbonate $[CuCO_3 \cdot Cu(OH)_2]$, which protects the rest of the metal from further action.

2. **Action of water :** Water at ordinary temperature has no action on copper. However, at white heat, steam acts upon the metal to give oxide.

$$Cu \quad + \quad H_2O \quad \longrightarrow \quad CuO + H_2 \uparrow$$

$$\text{(red hot)} \quad \text{(steam)}$$

3. **Action of acids :**

(a) **Action of hydrochloric acid :**

(i) Cold and dilute HCl has no action on copper.

(ii) Copper reacts with warm dilute HCl in the presence of oxygen to form cupric chloride.

$$2Cu + 4HCl + O_2 \quad \longrightarrow \quad 2CuCl_2 + 2H_2O$$

(iii) Copper powder dissolves slowly in hot and concentrated HCl with the evolution of hydrogen.

$$2Cu + 2HCl \quad \longrightarrow \quad Cu_2Cl_2 + H_2 \uparrow$$

(b) **Action of sulphuric acid :** Sulphuric acid behaves exactly in the same manner as that of hydrochloric acid.

(i) Cold and dil. H_2SO_4 has no action on copper.

(ii) It dissolves slowly in hot and dil. H_2SO_4 in the presence of oxygen.

$$2Cu + 2H_2SO_4 + O_2 \quad \rightarrow \quad 2CuSO_4 + 2H_2O$$

(iii) On heating with conc. H_2SO_4, it produces SO_2.

$$Cu + H_2SO_4 \quad \rightarrow \quad CuO + H_2O + SO_2$$

$$\underline{CuO + H_2SO_4 \quad \rightarrow \quad CuSO_4 + H_2O \qquad\qquad}$$

$$Cu + 2H_2SO_4 \quad \rightarrow \quad CuSO_4 + SO_2 + 2H_2O$$

(c) **Action of nitric acid :**

(i) Dilute HNO_3 dissolves the metal with the evolution of nitric oxide.

$$Cu + 2HNO_3 \quad \rightarrow \quad Cu(NO_3)_2 + 2H] \times 3$$

$$\underline{HNO_3 + 3H \quad \rightarrow \quad NO + 2H_2O] \times 2 \qquad\qquad}$$

$$3Cu + 8HNO_3 \quad \rightarrow \quad 3Cu(NO_3)_2 + 4H_2O + 2NO$$

(ii) Conc. HNO_3 dissolves the metal and gives nitrogen dioxide.

$$Cu + 2HNO_3 \quad \rightarrow \quad Cu(NO_3)_2 + 2H$$

$$\underline{2HNO_3 + 2H \quad \rightarrow \quad 2NO_2 + 2H_2O \qquad\qquad}$$

$$Cu + 4HNO_3 \quad \rightarrow \quad Cu(NO_3)_2 + 2H_2O + 2NO_2 \uparrow$$

(d) **Action of conc. HBr and HI :** Copper dissolves in these acids with the formation of complex salts e.g.

$$2Cu + 2HBr \quad \rightarrow \quad 2CuBr + H_2$$

$$2CuBr + 2HBr \quad \rightarrow \quad H_2[Cu_2Br_4]$$

(e) **Action of acetic acid :** In the presence of air, copper reacts with acetic acid to form cupric acetate.

$$Cu + 2CH_3COOH \quad \rightarrow \quad (CH_3COO)_2Cu + H_2O$$

(5) Uses of Copper

(1) For making electrical wires, cables and conducting apparatus.

(2) As a coinage metal and in ornaments, jewellery for making them hard.

(3) For making water stills, kettles, vacuum pans, steam pipes, and fire boxes of locomotive engines.

(4)   For making scientific apparatus such as hyposometers, colorimeters etc.

(5)   In electroplating and electrotyping.

(6)   Copper salts are largely used as insecticides and colouring materials.

(7)   Large quantities of copper are used for making various alloys like brass, bronze, gun metal etc.

2.2.6 Metallurgy of Aluminium

(1) Occurrence

- Aluminium occurs in combined state.

- It is the third most abundant element (8.31% by weight) after oxygen and silica.

- It occurs widely as constituents of rocks and soils.

- Its main ores are :

    (i)   Oxides : Bauxite ($Al_2O_3 \cdot 2H_2O$), Corundum (Al_2O_3).

    (ii)   Fluoride : Cryolite (Na_3AlF_6).

    (iii)   Silicates : Feldspar ($KAlSi_3O_8$), Mica ($KAlSi_2O_{10}(OH)_2$)

    (iv)   Basic sulphates : Alunite or alumstone $K_2SO_4 \cdot Al_2(SO_4)_3 \cdot 4Al(OH)_3$.

- Bauxite ore is commonly used to extract aluminium.

- In India, bauxite is found in M.P., Bihar, Maharashtra, Mysore, Orissa, Tamil Nadu and Kashmir.

(2) Extraction of Aluminium

- Aluminium is usually isolated from bauxite by electrolysis. Since it is difficult to purify aluminium, it is necessary to use pure bauxite during electrolysis.

- Extraction involves following three steps :

    I.   Purification of bauxite.

    II.   Electrolytic reduction of alumina.

    III. Refining of aluminium.

I.   Purification of bauxite :

- Bauxite usually contains oxides of iron and silica as impurities.

- Purification method varies according to the type of impurities present as follows :

    (A) Baeyer's process : For the removal of impurities of iron (or for purification of red bauxite).

    (B) Serpeck's process : For the removal of impurities of silica (or for purification of white bauxite).

    (C) Hall's process : For the removal of both iron and silica impurities.

(A) Baeyer's process :

- The powdered bauxite ore is roasted to convert ferrous oxide (FeO) to ferric oxide (Fe_2O_3).

- This roasted ore is then heated with concentrated NaOH (45%) under pressure for few hours in autoclave. This process is known as leaching.

- Aluminium oxide dissolves forming sodium meta aluminate, while ferric oxide remains undissolved.

$$Al_2O_3 + 2NaOH \xrightarrow[\text{80 atm.}]{\text{150°C}} 2NaAlO_2 + H_2O$$

sodium meta aluminate

Undissolved Fe_2O_3 is removed by filtration.

- The filtrate (containing sodium meta aluminate) is diluted with water to form a precipitate of aluminium hydroxide $[Al(OH)_3]$.

$$NaAlO_2 + 2H_2O \longrightarrow NaOH + Al(OH)_3 \downarrow$$

- The precipitate of $Al(OH)_3$ is then filtered out, dried and heated at 1500°C to get pure alumina.

$$2Al(OH)_3 \xrightarrow{\quad 1500°C \quad} Al_2O_3 + 3H_2O$$

(B) Serpeck's process :

- This method is suitable to purify silica containing bauxite (white bauxite).

- In this method, the powdered ore is mixed with carbon and heated to 1800°C in a current of nitrogen.

- Silica present as impurity is reduced to silicon, which volatilizes off while alumina forms aluminium nitride.

$$SiO_2 + C \rightarrow Si \uparrow + 2CO \uparrow$$

$$Al_2O_3 + N_2 + 3C \rightarrow 2AlN + 3CO \uparrow$$

$$\text{alumina} \qquad\qquad \text{aluminium nitride}$$

- Aluminium nitride is hydrolysed with water to give aluminium hydroxide precipitate. This precipitate is filtered out, dried and ignited to get pure alumina.

$$AlN + 3H_2O \longrightarrow Al(OH)_3 + NH_3$$

$$2Al(OH)_3 \xrightarrow{\quad 1500°C \quad} Al_2O_3 + 3H_2O$$

(C) Hall's process :

- This method is used for the ore having both Fe_2O_3 and SiO_2 as impurities.

- The powdered ore is heated to bright redness with sodium carbonate. Aluminium oxide, being amphoteric, dissolves to form sodium meta aluminate while the insoluble Fe_2O_3 and SiO_2 are left as residue.

$$Al_2O_3 \cdot 2H_2O + Na_2CO_3 \longrightarrow 2NaAlO_2 + 2H_2O + CO_2$$

- The fused mass is extracted with water and the insoluble Fe_2O_3 and SiO_2 are removed by filtration.

- The filtrate (water extract) is heated to 50-60°C and a current of CO_2 is passed through it when $Al(OH)_3$ is precipitated out.

$$2NaAlO_2 + 3H_2O + CO_2 \longrightarrow 2Al(OH)_3 \downarrow + Na_2CO_3$$

- $Al(OH)_3$ is filtered off, dried and ignited to get pure alumina.

II. Electrolytic reduction of alumina :

- Alumina cannot be reduced to metallic aluminium by carbon, because aluminium has great affinity for oxygen and thus the reduction of alumina (Al_2O_3) by carbon (smelting) under ordinary conditions is not possible.

  Further at high temperature, aluminium reacts with carbon to form aluminium carbide (Al_4C_3).

- Electrolytic reduction of alumina also has the following difficulties :

  (i) Pure alumina has very high melting point (2000°C). In case the electrolysis is carried out at 2000°C, aluminium obtained vaporises because its melting point is 1800°C.

  (ii) Alumina is a bad conductor of electricity.

- Hall and Heroult overcome these difficulties by carrying the electrolysis of alumina in the presence of cryolite (Na_3AlF_6), which acts as a proper electrolyte.

- The purified alumina as obtained above is dissolved in molten cryolite (Na_3AlF_6) and is electrolysed in an iron tank lined inside with carbon, which serves as a cathode. (Cryolite decreases the melting point of alumina and also increases the electrical conductivity of the electrolyte.)

- The anode is made up of carbon rods attached to copper clamps and suspended in the fused electrolyte. The carbon rods are so arranged that these can be raised, lowered or replaced as desired.

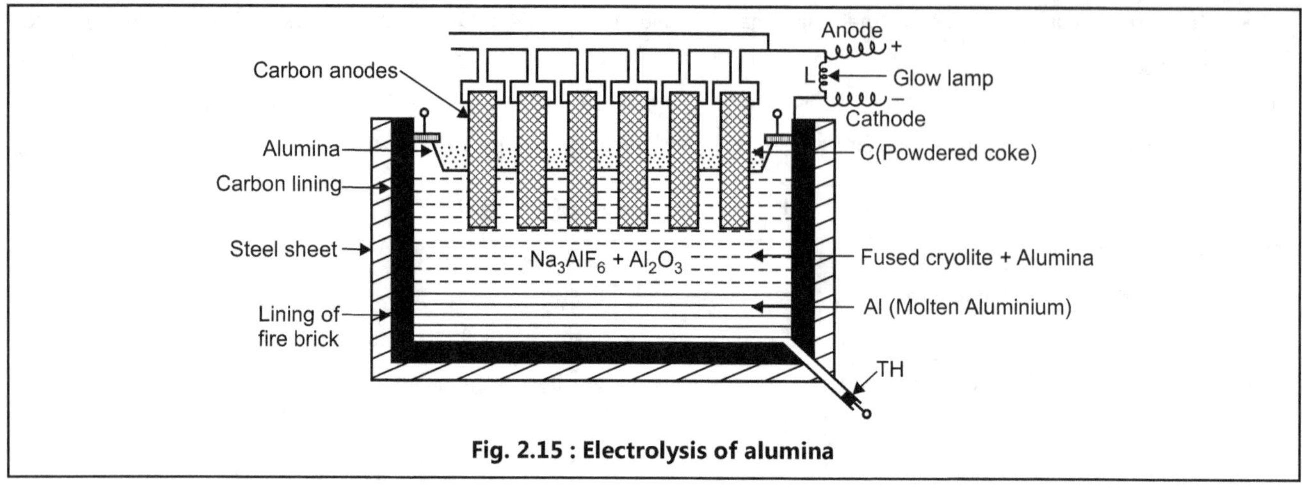

Fig. 2.15 : Electrolysis of alumina

- Electrolytic bath is covered with powdered coke to avoid the oxidation of metal and to avoid the loss of heat.

- Heat produced by the current keeps the mass in fusible state. The temperature of bath is maintained at 900-1000°C.

- On passing current, alumina decomposes to aluminium and oxygen.

$$2Al_2O_3 \longrightarrow 4Al + 3O_2$$

- The molten aluminium sinks to the bottom (cathode) from where it is tapped off from time to time. The metal obtained is about 99% pure. While oxygen is liberated at anode, which then combines with carbon anodes and form CO and CO_2. These gases are allowed to escape through the outlets.

- Since carbon anodes are constantly consumed due to formation of CO and CO_2, they are required to be replaced from time to time.

- After a certain time, the concentration of alumina falls in the cell, below a certain limit, the resistance of the cell increases and therefore, more current flows through the control lamp connected in parallel with the cell and hence it glows up. This indicates that alumina is exhausted and hence more alumina should be added to continue the process.

- The exact nature of the reaction taking place in the cell is very complicated and not properly understood. Most probably, it is as shown below.

$$Na_3AlF_6 \rightleftharpoons 3NaF + AlF_3$$

cryolite

$$AlF_3 \rightleftharpoons Al^{+3} + 3F^-$$

$$\text{(at cathode)} \quad \text{(at anode)}$$

At cathode : $Al^{+3} + 3e^- \rightarrow Al$

At anode : $F^- \rightarrow F + e^-$

$$2Al_2O_3 + 12F \rightarrow 4AlF_3 + 3O_2 \uparrow$$

Liberated O_2 combines with carbon anodes to form CO and CO_2.

$$2C + O_2 \rightarrow 2CO \uparrow$$

$$2CO + O_2 \rightarrow 2CO_2 \uparrow$$

- Thus, cryolite is used only for starting the reaction, because fluorine atoms from cryolite convert Al_2O_3 to AlF_3 which ionises and the reactions go on till the alumina is exhausted.

III. Refining of aluminium (Hoope's process) :

- Crude aluminium is 90% pure. It contains small quantities of Fe, Si, C and Al_2O_3. It can be further purified by Hoope's process.

- The electrolytic cell consists of an iron tank lined inside with carbon.

- The tank contains three layers of fused masses one over the other according to their densities.

(i)   The bottom layer consists of impure molten Al which along with carbon lining acts as the anode.

(ii)   The middle layer consists of mixture of molten fluorides of Na, Al and Ba. It acts as a electrolyte.

(iii)   The top layer consists of pure aluminium having carbon electrodes dipped in it. It acts as an cathode.

- On passing electric current, aluminium ions from the middle layer are discharged at the cathode as pure aluminium. The pure aluminium is removed from the taping hole.

- An equivalent amount of aluminium from the bottom layer moves into the middle layer leaving behind the impurities.

- This method gives 99.99% pure aluminium.

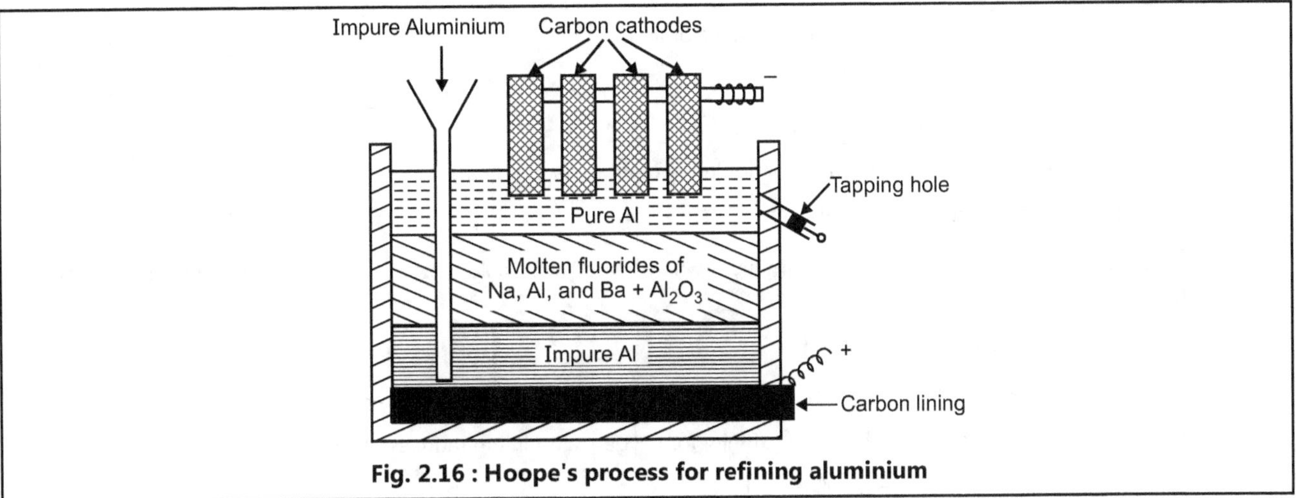

Fig. 2.16 : Hoope's process for refining aluminium

3. Properties of Aluminium

[A] Physical properties :

(1)   Aluminium is a silvery white metal with a brilliant lustre which is soon destroyed due to the formation of oxide layer. It takes a very high polish.

(2)   It is a very light metal having sp. gr. 2.7.

(3)   It is malleable and ductile and can be rolled into sheets, foils and wires.

(4)   It is a good conductor of heat and electricity.

(5)   It melts at 659°C and boils at 2450°C.

[B] Chemical properties :

1.   Action of air : Aluminium is not affected in dry air, but in moist air a thin film of oxide is formed over its surface which protects the metal from further corrosion. Thus, aluminium is said to be a self-protective metal.

On heating in air or oxygen, it burns readily producing brilliant light with the evolution of a huge quantity of heat.

$$4Al + 3O_2 \rightarrow 2Al_2O_3 + 772,000 \text{ cal}$$

2.   Action of water : Pure aluminium is not affected by pure water. However, the commercial (impure) aluminium is readily corroded by water containing salts (sea water). It decomposes boiling water with the evolution of hydrogen.

$$2Al + 6H_2O \rightarrow 2Al(OH)_3 + 3H_2 \uparrow$$

Amalgamated aluminium (Al-Hg) reacts most instantly with cold water liberating hydrogen. Hence, it is used as a reducing agent.

3.   Action of acids : Aluminium dissolves readily in HCl to form aluminium chloride and liberates hydrogen.

$$2Al + 6HCl \rightarrow 2AlCl_3 + 3H_2 \uparrow$$

Dil. H_2SO_4 reacts slowly with aluminium and liberates hydrogen.

$$2Al + 3H_2SO_4 \rightarrow Al_2(SO_4)_3 + 3H_2 \uparrow$$

Hot and conc. H_2SO_4 liberates SO_2 gas with aluminium.

$$2Al + 6H_2SO_4 \rightarrow Al_2(SO_4)_3 + 6H_2O + 3SO_2 \uparrow$$

Nitric acid (dil. and conc.) has no action on pure aluminium.

4.   Action of alkalies : Aluminium readily dissolves in strong alkalies like NaOH, KOH forming metal aluminates or aluminates and liberating H_2 gas.

$$2Al + 2NaOH + 2H_2O \rightarrow 2NaAlO_2 + 3H_2 \uparrow$$

sodium meta aluminate

$$2Al + 6NaOH \rightarrow 2Na_3AlO_3 + 3H_2 \uparrow$$

sodium aluminate

It also dissolves in hot conc. solution of Na_2CO_3 forming sodium meta aluminate.

$$2Al + Na_2CO_3 + 3H_2O \rightarrow 2NaAlO_2 + CO_2 + 3H_2$$

5.   Action of halogens and nitrogen : Aluminium when heated with halogens and nitrogen, forms halides and nitrides respectively.

$$2Al + 3Cl_2 \rightarrow 2AlCl_3$$

$$2Al + N_2 \rightarrow 2AlN$$

6.   Reducing action : On account of its great affinity for oxygen, aluminium reduces many metallic oxides.

$$Fe_2O_3 + 2Al \rightarrow Al_2O_3 + 2Fe$$

$$Cr_2O_3 + 2Al \rightarrow Al_2O_3 + 2Cr$$

7.   Displacement of other metals : Aluminium displaces copper, zinc, lead from their salt solutions.

$$3ZnSO_4 + 2Al \rightarrow Al_2(SO_4)_3 + 3Zn$$

4.   Uses of Aluminium

- Aluminium, being a very light metal, is used in household utensils, aeroplane parts, precision and surgical instruments.

- Being light and a good conductor of electricity, it is used to some extent in place of copper, in electric wires and cables for transmission lines.

- Aluminium forms foils of extreme thinness, which are used as wrappers for confectioneries (like chocolates) and cigarettes.

- Since it is unaffected by HNO_3, it is used in chemical plants and also for transporting nitric acid.

- Aluminium powder mixed with linseed oil is used in paints.

- Aluminium is largely used as a refractory for lining of furnaces and for making refractory bricks.

- It is used for making mirrors in telescope and optical instruments.

- A mixture of aluminium powder and NH_4NO_3 is used as an explosive in bombs under the name 'ammonal'.

- Amalgamated aluminium (Al-Hg) is used as a reducing agent.

- Due to its great affinity for oxygen, it is used for the extraction of metals (like Fe, Cr, Mn etc.) from their oxides (aluminothermic process) and in thermite welding.

- Its alloys are used in making parts of air-crafts, automobiles and steel boats.

- Aluminium powder is used in flashlight bulbs for indoor photography.

(B) ALLOYS

2.3 ALLOYS

2.3.1 Definition of Alloys

- In nature, it is difficult to get any substance in its pure state. This is true for the metals also. When we extract metals from their natural sources - minerals or ores - they contain some impurities. Hence these metals have to be purified. Various methods are used for purification of metals. But when we get the pure metal it is found that it becomes practically useless for engineering uses. These pure metals have very few useful properties like metallic lustre, good electrical conductivity, high malleability and ductility. The pure metals are very soft, highly chemically reactive. These properties like softness and high malleability reduce their shock and wear resistance. The high chemical reactivity makes the pure metals susceptible for corrosion and therefore they cannot be used for engineering purposes. Iron is used in the form of steel which shows the desired properties such as hardness, toughness, corrosion resistance etc. The steel is nothing but an alloy of iron with carbon, nickel, chromium, manganese etc.

2.3.2 Properties of Alloys

- The properties of a given metal can be improved by alloying it with some other metals or non-metals like carbon, phosphorus. When two or more metals are mixed in their molten state, they solidify without separation. On solidification they form a homogeneous substance called an alloy. Thus an alloy is a substance formed by solidification of a metallic solution. An alloy can also be defined as a *"solid solution where the solutes are the alloying elements and the solvent is the element in excess proportion."* In alloys, the chemical properties of constituent elements are retained while physical properties are improved.

- An alloy is a homogeneous mixture of two or more metals. In some alloys, non-metals like carbon, boron, phosphorus, sulphur can be mixed with metals. For example, plain carbon steel is an alloy of iron and carbon; phospher bronze contains copper, tin and phosphorus. *Hence an alloy is to be defined as a homogeneous mixture of two or more elements one of which must be a metal.*

- When alloy contains mercury as one of the components then it is called *amalgam, e.g. sodium amalgam* (Na - Hg), aluminium amalgam (Al - Hg), zinc amalgam (Zn - Hg) etc.

2.3.3 Preparation of Alloys

- An alloy can be prepared by any of the following methods :

    (1) Fusion, (2) Electro-deposition, (3) Compression, (4) Reduction.

    (1) Fusion method : In this method, the component metal having higher melting point is melted first in a crucible and the other components having lower melting points are added to it in the required quantity. In the manufacture of alloys, the molten metals are at higher temperature and hence react with atmospheric oxygen to form oxide. This oxidation is prevented by covering the molten mass with a fine charcoal powder. To get a uniform alloy, the molten mixture is stirred using graphite rods. The specific gravities of constituents have also to be considered. The heavy metals are generally mixed at the end to avoid its settling due to gravity which gives uneven composition in alloys. The molten mass is then allowed to cool which gives the required alloy.

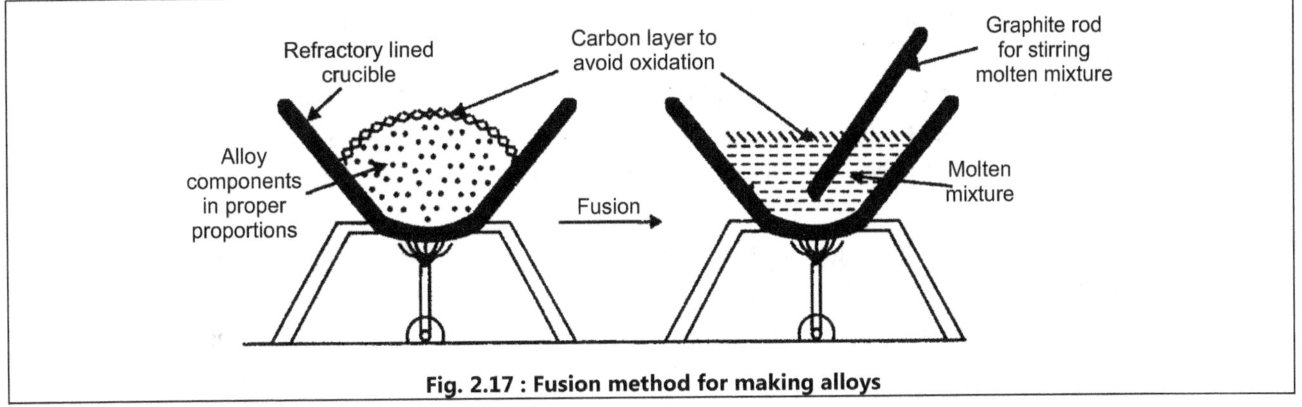

Fig. 2.17 : Fusion method for making alloys

In the manufacture of brass, an alloy of copper and zinc, the copper (melting point 1089°C) is melted first and the required quantity of zinc (melting point 419°C) is added to it which melts immediately. The molten mass is stirred and allowed to cool. To avoid oxidation of copper and zinc, the surface of molten mass is covered with charcoal powder. The mass is allowed to cool slowly to get brass.

Manufacture of bronze, an alloy of copper and tin, is exactly similar to the above process. In this case, requisite quantity of tin (m.p. 232°C) is added to the molten copper (m.p. 1089°C) to get bronze.

(2) Electro-deposition : The alloys can be prepared by simultaneous electro-deposition of the components from the electrolyte containing their salt solution. An alloy, brass of copper and zinc can be obtained by this method. The electrolysis of a mixed solution of copper and zinc cyanides dissolved in potassium cyanide is carried out to obtain brass.

(3) Compression : The alloy is made by intimately mixing two or more metal powders and the intimate mixture is compressed under a high pressure in a mould. The moulded article is then heated to a temperature just below the melting point of an alloy. Due to this, the tiny particles of metals are firmly welded to one another. *Solder alloy,* an alloy of lead and tin, is obtained by this method. An alloy known as *Wood's metal* containing lead, tin, bismuth and cadmium is similarly obtained by this method.

(4) Reduction : The alloy is obtained by the reduction of a suitable compound, generally oxide of one component metal in the presence of the other component metal. The component metal oxide is reduced to metal in the presence of other metal. For example, aluminium bronze is prepared by reducing aluminium oxide in the presence of copper in an electric furnace.

2.3.4 Purposes of Making Alloys

- Metals are used as a basic material in the manufacture of machinery, household articles, ships, railways, aircrafts, bridges, buildings etc. The properties of one metal do not fulfil all the service requirements. The mechanical properties such as tensile strength, malleability, ductility, elastic limit, hardness are to be modified to make the metal suitable for a given purpose. This modification in the properties of the metal can be effected by addition of one or more metals or elements to the original metal to form an alloy. By making alloys the required properties can be developed. The deficiency of one metal can be corrected by the addition of one or more metals and while doing so care is taken to see that the valuable properties of original metal are not lost.

  The main purposes of alloying are :

  (1) To improve the hardness of metal.

  (2) To lower the melting point.

  (3) To increase the tensile strength.

  (4) To increase corrosion resistance.

  (5) To get good casting.

  (6) To modify the colour.

  (7) To reduce the malleability and ductility.

  (8) To modify chemical activity.

- The metal which constitutes the major portion of the alloy is known as base metal and others are called as alloying metals or elements.

  The general improvements in properties of metals which can be obtained by alloying are as follows :

  (1) To improve the hardness of metal : Pure metals are generally soft. But when the metal is alloyed with another metal or non-metal, its hardness is increased. *"An alloy is harder than its component metals."*

For example :

  (1) Pure gold and silver are soft. Hence they are hardened by the addition of a small amount of copper while using them in ornaments and coins. The addition of small amount of copper makes them hard enough to resist wear and tear.

  (2) Pure iron is very soft and cannot be used as such for making machinery parts. Hence the iron has to be converted to steel by the addition of small quantity (0.05 - 1.5 %) of carbon. Thus, presence of carbon imparts hardness to steel.

(3) Brass (an alloy of Cu and Zn) and bronze (an alloy of Cu and Sn), both are harder than the base metal copper.

(4) Lead is soft metal but its hardness can be improved by the addition of 0.5 % arsenic to it. This alloy is used for making lead shots.

(2) To lower the melting point : When an alloying element is added to a base metal, it acts as an impurity in base metal and the melting point of base metal is lowered. *"In general, the melting point of an alloy is lower than those of its constituent elements."* Thus, alloying makes the metal easily fusible. This property of the metal is used in making useful low melting alloys called *solders*. The solder metal must have lower melting point than the metal to be soldered and should contain the same metal as one of its constituents.

For example :

(1) Wood's metal is an alloy of bismuth, lead, tin, and cadmium. It has the melting point of only 71°C, which is much lower than those of its components Bi, Pb, Sn, and Cd.

(2) Rose metal is an alloy of bismuth, lead and tin having the melting point 100°C. Other alloys which are easily fusible are Newton metal, Solder alloy etc.

(3) To increase the tensile strength : When a metal is alloyed with proper elements (like Ni, Cr, V etc.), the elasticity and tensile strength is adjusted to the requirement i.e. the tensile strength of pure metal is increased by alloying.

For example :

(1) The addition of 1 % carbon increases the tensile strength of pure iron by about ten times.

(2) The addition of 1 % nickel or chromium to mild steel increases its tensile strength tremendously. The tensile strength of steel is further increased by the addition of 0.15 % vanadium and 1 % chromium to it.

(3) Tensile strength of copper can be doubled by adding about 5 % of silicon to it.

(4) To increase corrosion resistance : The metals in pure form (like Fe and Cu) are quite reactive and are easily corroded by surrounding atmospheric gases, moisture, etc., thereby their life is reduced. But if a metal is alloyed, it resists corrosion. In other words, *"alloys are more resistant to corrosion than pure metals"*.

For example :

(1) Pure iron is corroded even in moist air, but its alloy like stainless steel (an alloy of Fe - C - Cr - Ni) does not even get stained i.e. it is totally resistant to corrosion.

(2) Bronze (an alloy of Cu - Sn) is more corrosion resistant than copper.

(3) Zinc is readily attacked by sulphuric acid, but brass (an alloy of Cu - Zn) is not attacked.

(4) The alloys of copper such as Naval brass (Cu - Zn - Sn) and German silver (Cu - Zn - Ni) are non-corrosive.

(5) To get good castings : To get good castings from a metal or alloy, it is essential that the metal must expand on solidification and it should be easily fusible so that sharp impression can be taken easily. Generally, pure molten metals contract on solidification. Hence, in order to get good castings, metals have to be alloyed, because alloys expand on solidification.

For example :

(1) Type metal (an alloy of lead, tin and antimony), which is used for casting types of printing due to its exceptionally good casting properties. Small addition of antimony gives the property of expanding on solidification while tin gives hardness.

(2) Bronze (an alloy of Cu-Sn) and duralumin (an alloy of Al-Cu-Mg-Mn) possess good casting properties.

(3) The casting property of aluminium can be improved by the addition of small amount of copper or magnesium to it.

(4) Gun metal (Cu - 88%, Sn - 10%, Zn - 2%) and phosphor bronze (Cu - 96%, Sn - 3.75%, P – 0.25 %) are the other alloys of copper having good casting properties.

(6) To modify colour : The colour of the metal is its characteristic property, hence we cannot change the colour of a given metal. But the colour of a metal can be modified to a desired one by alloying it with another suitable element. Thus an alloy can be prepared having colour quite different from the colour of the base metal.

For example :

(1) Both aluminium and tin are silvery white in colour, but their alloy, aluminium bronze (Cu-Al-Sn) has a beautiful golden colour.

(2) Brass is an alloy of copper (red) and zinc (silvery white) and is yellow in colour.

(3) Silver is white, but alloy of silver and tin has pink colour, similarly, an alloy of gold and silver is purple.

(7) To reduce malleability and ductility : Pure metals are highly malleable and ductile which results in lowering the toughness. The shape of articles prepared from pure metals get changed even due to small force. To increase resistance of metal to such forces i.e. to make it tough it is necessary to reduce its malleability or ductility which is effected by alloying with some suitable metal.

For example : A small amount of copper is added to gold and silver to reduce their malleability and ductility.

(8) To modify chemical activity : The chemical reactivity of a metal can be changed by alloying it with other metals. This does not affect the products of reaction but changes the rate of reaction.

For example :

(1) Sodium is a highly reactive element, but when it is alloyed with mercury to form an alloy called sodium-amalgam (Na-Hg), it becomes less reactive.

(2) The chemical reactivity of aluminium increases when it is alloyed with mercury to form aluminium-amalgam (Al-Hg).

2.3.5 Classification of Alloys

Alloys are generally classified into two classes :

(1) Ferrous alloys : The alloys containing iron as one of the main components are known as ferrous alloys. e.g. alloy steels are the most common ferrous alloys.

(2) Non-ferrous alloys : The alloys which do not have iron as one of the main components are known as non-ferrous alloys. e.g. brass, bronze (alloys of copper), duralumin (alloy of aluminium) are some of the common non-ferrous alloys.

2.4 FERROUS ALLOYS (Nov. 18)

2.4.1 Various Methods of Steel Making (Nov. 18)

Following processes are carried out for the conversion of cast iron into steel :

    (i) Bessemer process.

    (ii) L.D. process.

(Manufacture of Steel): (Conversion of Cast iron into Steel)

In general, steel is manufactured by blowing hot air or oxygen through molten cast iron whereby impurities are oxidized and removed as volatile gases or subsequent slags. The process is then followed by adding precise small quantities of carbon to the pure molten iron; steel of required strength and malleability is obtained. At present, steel-making is done by several different methods.

(A) Old processes : (Hot air is blown)

    1. Electric process,

    2. Bessemer process,

    3. Open-hearth process,

    4. Duplex process, etc.

(B) Modern processes: Oxygen is blown: Basic oxygen process (BOP)

1. Kaldo process
2. Linz – Donawitz process

1. Bessemer Process:

This process is invented in 1855 by Henry Bessemer (a Frenchman, Resident of England.) In this method, a furnace called Bessemer converter is used. It is a pear or egg shaped vessel constructed by using steel plates. It is a pear or egg shaped vessel constructed by using steel plates. It is supported on side arms called trunnions, whereby it can be oriented in different angles:

(i)   It is tilted to horizontal position for charging,

(ii)   It is inverted to pour out the finished product, and

(iii)   It is kept vertical to run the process of oxidation. (Refer Fig. 2.18)

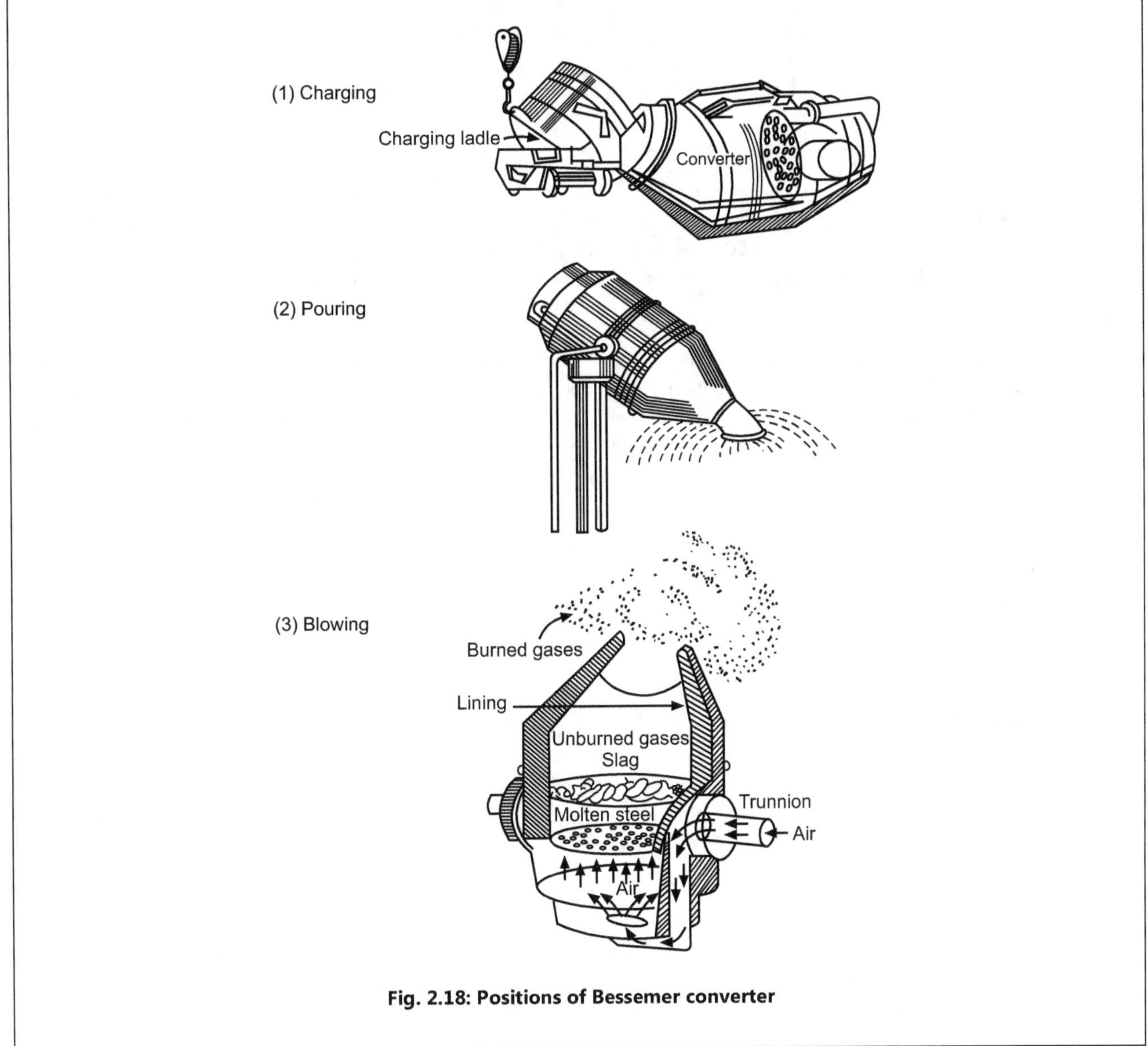

Fig. 2.18: Positions of Bessemer converter

It is provided with twyers or nozzles at the bottom to blow hot air. Hence it is called bottom-blown converter. (Refer Fig. 2.19)

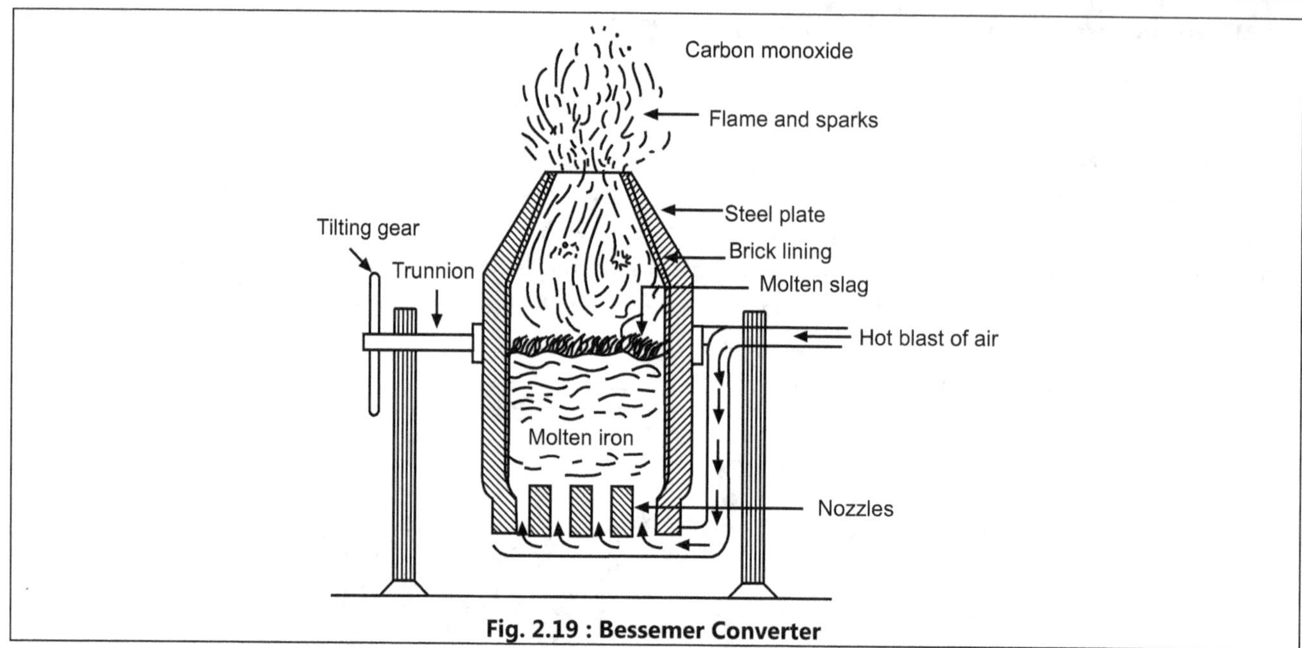

Fig. 2.19 : Bessemer Converter

From inside it is lined with refractory lime bricks or silica bricks. When acidic impurities are to be removed (such as SiO_2, P_4O_{10}), a basic lining such as dolomite ($CaCO_3 . MgCO_3$) is used. This process is called Basic Bessemer Process. When basic impurities are to be removed, acidic lining of silica bricks (SiO_2) is used and the process is called *Acid Bessemer process.*

Working:

(a) **Acid Bessemer Process:** In this process the converter is lined with silicious refractory material which is acidic in nature. To start, the converter is turned into the horizontal position, and charged with about 60 tonnes of molten cast iron, at about (1200°C). It is turned to vertical position and a hot blast of air is blown from the bottom. Impurities get oxidized and temperature rises to 2173 K (1900°C). The various reactions take place in the converter.

$$2Mn + O_2 \longrightarrow 2MnO$$

$$Si + O_2 \longrightarrow SiO_2$$

$$2C + O_2 \longrightarrow 2CO$$

$$S + O_2 \longrightarrow SO_2$$

The oxides of silicon and manganese combine to form a slag of manganese silicate.

$$MnO + SiO_2 \longrightarrow MnSiO_3 \text{ (slag)}$$

A little amount of iron is also oxidized to ferric oxide but it gets readily reduced by carbon present in the cast iron.

$$4Fe + 3O_2 \longrightarrow 2Fe_2O_3$$

$$Fe_2O_3 + 3C \longrightarrow 2Fe + 3CO$$

Ferric oxide formed above also oxidises manganese and silicon to their respective oxides.

$$3Mn + Fe_2O_3 \longrightarrow 3MnO + 2Fe$$

$$3Si + 2Fe_2O_3 \longrightarrow 3SiO_2 + 4 Fe$$

The course of the reaction in the plant is followed from the burning of waste gases and observing the colour of the flame produced. The carbon monoxide produced in the above reactions burns at the mouth of the converter with a blue flame and orange-red tinge, throwing out showers of sparks. When the whole of carbon is oxidized, the blue flame suddenly dies down.

When blue flame at the top dies, a calculated quantity of Spiegeleisen (alloy of Fe, Mn and C) is added followed by the addition of scavenger such as ferrosilicon or aluminium. Blast of air is passed, mass is mixed well, slag is tapped off, the converter is tilted down and molten steel is taken out.

Role of Spiegeleisen:

Carbon and Mn present in it reduce FeO from the molten iron.

$$FeO + C \longrightarrow Fe + CO$$

$$FeO + Mn \longrightarrow Fe + MnO$$

It contains 20 to 33% Mn and about 6% carbon. Here carbon regulates the desired quality of steel.

Role of Scavenger:

The molten iron contains dissolved gases such as O_2, N_2 and CO_2 etc. which create blow holes or gas bubbles in castings and defects are created. When aluminium is added, it reacts with these gases and removes them as slag.

$$2Al + N_2 \longrightarrow 2AlN$$

$$4\,Al + 3O_2 \longrightarrow 2Al_2O_3$$

Thus spiegeleisen acts as a deoxidizer while Al acts as a scavenger.

(b)   Basic Bessemer (Thomas-Gilchrist Process): This process is used to treat cast iron containing phosphorus. In this process, the converter is lined with magnesia and lime prepared by calcinations of dolomite ($CaCO_3.MgCO_3$). Some limestone is added into the converter. Molten cast iron from blast furnace is then run into the converter and the blast continued. Carbon, sulphur and manganese are oxidized first as usual, but if the blast is continued even after the flame sinks down, then the phosphorus forms phosphorus pentoxide.

$$P_4 + 5O_2 \longrightarrow P_4O_{10}$$

The phosphorus pentoxide thus formed, combines with lime to form a basic slag, containing calcium phosphate.

$$6CaO + P_4O_{10} \longrightarrow 2Ca_3(PO_4)_2 \text{ (slag)}$$

The slag, also known as Thomas slag, is used as a valuable fertilizer. As in Acid Bessemer Process, after complete removal of carbon and other impurities, requisite amount of carbon in the form of spiegeleisen is added and the product thoroughly mixed by air-blow.

In recent plants instead of air, oxygen diluted either with steam or carbon dioxide is used. It is found that nitrogen from air is retained in steel and makes it brittle. Pure oxygen is avoided as it burns the nozzles very quickly and further a large volume of iron oxide fumes are formed. Further CO_2 is preferred to steam, since hydrogen in the later, tends to make the steel brittle.

Merits and Demerits of Bessemer Process:

1. The process is useful for rapid production of steel.
2. Since the molten cast iron obtained from blast furnace, is directly taken in the converter, no extra fuel is required for melting of cast iron.
3. Hence, the cost of operating the process is low.
4. But it should be noted that the process is not continuous and charging is rather tedious.
5. Comparatively small quantities (about 5 to 8 tonnes) of iron can be converted into steel, in 20 minutes, in conventional Bessemer converters.
6. This steel produced is of inferior quality.
7. The loss of iron in the slag is comparatively more (about 15%).

2.   L.D. Process (Linz Donawitz Process):

This is the most recent process due to Linzer and Dusen Verfarhen. It was first operated in 1953 in Austria at Linz-Donawitz and therefore carries the name L.D. Process. The furnace is an egg shaped steel vessel, sealed from bottom and supported on the side arms called trunnions. It is lined from inside with basic lining of magnesite or lime or limestone ($CaCO_3$). At the top, it is provided with water-cooled copper lance. Pure and dry oxygen is passed under pressure of 10 atm. (10×10^5 Pa) through the copper lance into the furnace. It is commonly known as L.D. converter and may be termed as a top-blown converter. (Refer Fig. 2.20).

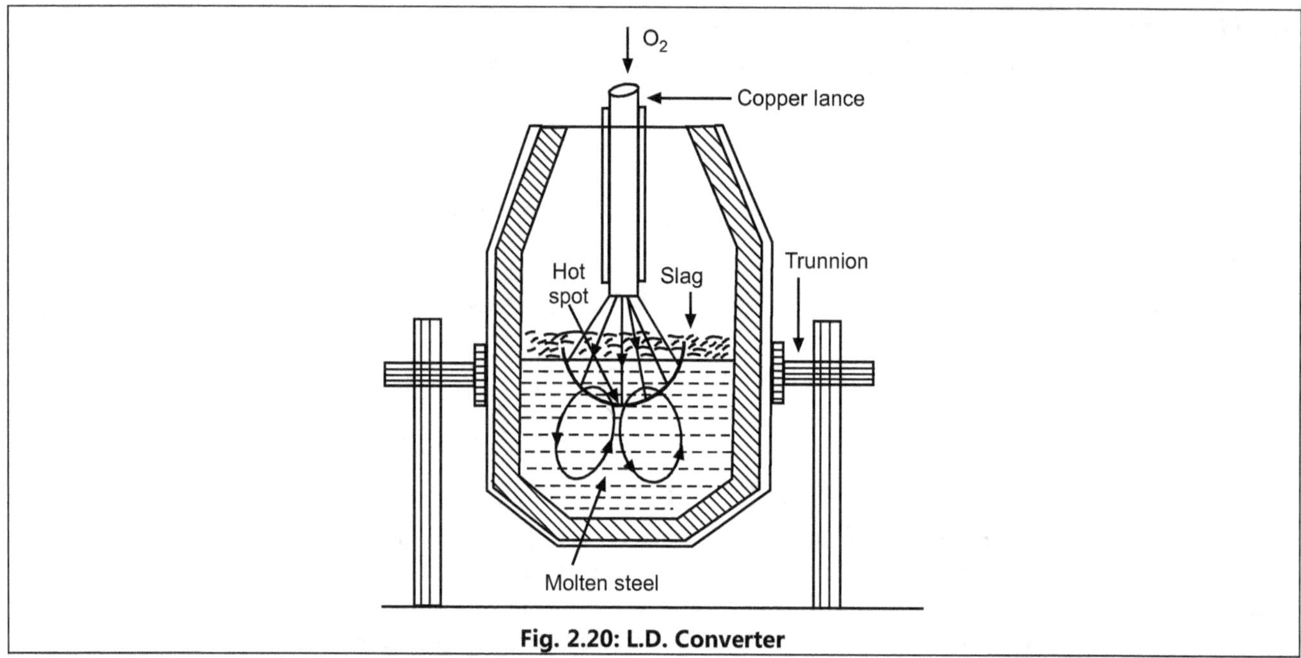

Fig. 2.20: L.D. Converter

Working: The converter is charged with cast iron (80%), scrap steel (18%) and limestone (2%). Pure and dry oxygen say (99.5 to 99.7%) is injected into the molten mass through copper lance as shown in Fig. 2.20. Oxygen jumps on the molten mass from the copper lance at 40 to 50 cm. The impurities like S, P, Mn, Si etc. are oxidized by oxygen and removed as slag by combining with lime. Very much heat is given out and temperature reaches to about (2773 K) 2500°C. Thus a hot spot is created at the top of charge. When metal is purified, it becomes dense and sinks to the bottom of the converter while the lighter and impure mass moves towards the hot spot. Thus convection currents are set in and entire metal gets purified.

$$2C + O_2 \longrightarrow 2CO$$
$$2Mn + O_2 \longrightarrow 2MnO$$
$$Si + O_2 \longrightarrow SiO_2$$
$$P_4 + 5\,O_2 \longrightarrow P_4O_{10}$$
$$MnO + SiO_2 \longrightarrow MnSiO_3 \text{ (Slag)}$$
$$CaO + SiO_2 \longrightarrow CaSiO_3 \text{ (Slag)}$$
$$6CaO + P_4O_{10} \longrightarrow 2Ca_3(PO_4)_2 \text{ (Thomas Slag)}$$
$$\text{Flux} \quad \text{Impurity}$$

When purification is complete, convection currents are stopped and the process is regarded to be completed. This takes about 45 minutes for one complete cycle and steel produced is about 30 to 60 tonnes. Now-a-days even larger quantities can be handled in a shorter time, say 300 tonnes in 40 minutes. L.D. Process or Basic Oxygen process (BOP) is much suitable to produce low carbon/mild steels. Hindustan Steel Ltd., Rourkela produces (75%) steel by this method.

Advantages:

1. L.D. process is very rapid. It requires about 45 minutes for one batch.
2. Its productivity is comparatively high; 30 to 60 tonnes in about 45 minutes, or even higher say 300 tonnes in 40 minutes in recent plants.
3. On unit cost basis, its capital expenditure is less.
4. The manufacturing cost is low.
5. Energy needed is comparatively low.
6. Steel is of superior quality.
7. Molten cast iron and scrap steel can be used directly.
8. Both carbon and phosphorus are removed simultaneously due to intense heat.
9. The process gives a purer product and the surface is free from nitride.
10. At present L.D. process is the most superior to all the processes known.

2.4.2 Composition, Properties and Applications of Plain Carbon Steel

Definition of plain carbon steels :

"It is the alloys of iron containing carbon upto 1.5%, either in combined state (Fe$_3$C) or as solid solution, are called steels." Hence steels are essentially "carbon steels" or "plain carbon steels."

Types of Steels:

According to the presence of alloying elements, steels are broadly classified into three categories:

(a)   (Plain) Carbon Steels, (b) Alloy Steels, and (c) Brittle Steels.

Carbon Steels:

Based on the percentage of carbon, steels are classified into following types:

(i)   Low Carbon Steel (Mild Steels): Steel containing 0.05 to 0.3% carbon. Mild steel is very soft, tough, malleable and ductile.

Use: It is used for thin soft wires, wires for ropes, nails, chains, tubes, rivets, bolts, gears, fan blades, girders, etc.

(ii)   Medium Carbon Steel (Structural Steel): Steel containing 0.3 to 0.6% carbon. Medium carbon steel has high tensile strength. It is shock resistant and has a good heat response.

Use: It is mainly used in railway engineering say for rail roads, wheels, railway axles and fish plates, crane hooks, crank shafts, rotors, etc. It is also used for machine parts, springs, etc.

(iii)   High Carbon Steel: Steel containing 0.6 to 0.9% carbon. High carbon steel is hard, tough, and resistant to wear and characterized by higher tensile strength.

Use: It is used for making tools, drills, knives, punches, hammers, etc.

(iv)   Higher Carbon Steel (Tool Steel): Steel containing 0.9 to 1.5% carbon. Tool steels are higher quality high-carbon steels. These can produce keen cutting edges.

Use: It is specifically used for preparing cutting tools, lathe tools, wood working tools, blades, razors, cutters, saws, knives, etc.

2.4.3 Effect of Various Alloying Elements on Steel

Alloy Steels (Special Steels):

Steel containing small quantities of transition elements, viz. nickel, cobalt, chromium vanadium, molybdenum, tungsten, etc. are called alloy steels or special steels. Such steels have wide range of special properties and thus find wide spread applications for number of purposes.

Alloy steels may be classified on the basis such as:

(a)   Chemical composition,

(b)   Structure, and

(c)   Purpose of application.

Some important alloy steels have been listed in Table 2.1 for brief information.

Table 2.1: Alloy Steel

	Name	Composition	Properties	Uses
1.	Molybdenum steel	0.3 to 3% Mo	Resistant to corrosion even at high temperature	Axles and cutting tools.
2.	Nickel steel platinate	46% Ni	Hard, tough, elastic and rustless. High coefficient of expansion equal to that of glass	Wire scales, electric bulbs, radio valves, etc.
3.	Chrome steel	2 to 4% Cr	High tensile strength, hard.	Ball bearings, Cutting tools, etc.
4.	Manganese steels	6 to 15% Mn	Very tough and hard, resistant to water.	Crushing machine, helmets, railroad, railway points, etc.
5.	Stainless steel	12 to 20% Cr, 8% Ni	Resistant to corrosion non-magnetic, bright shining, stainless.	Ornamental units, cutlery, utensils, surgical appliances, chemical plants, etc.
6.	Nickel steel	2.5-5% Ni	Very hard, tough, rustless and elastic.	Wires, cables, gears, aeroplane parts, crankshafts, guns, etc.
7.	Cobalt steel	Upto 35% Co	Very hard, resistant to corrosion, very high magne-tism.	High speed tools and permanent magnets.
8.	Silicon steel	Upto 15% Si	Extremely hard, resistant to acids.	Pumps, pipes carrying acids, electromagnets and transformers.
9.	Tungsten steel	15-20% W, ~5% Cr, a little V	Retains hardness even at high tempe-rature	High speed machine and drilling tools, cutting tools, etc.
10.	Invar	~35% Ni with ~Mn	Hard, resistant to corrosion, very low coeffi-cient of thermal expansion.	Pendulum rods, balance wheels, etc.

Brittle Steels:

Steel or iron containing excessive amounts (i.e. more than 0.06%) of phosphorus and sulphur becomes brittle. Sulphur causes brittleness in hot while phosphorus makes it brittle in cold (i.e. red short and cold short respectively).

Conclusion: Presence of different elements and their different extents alter the properties of steel radically. Steel is mainly the alloy of iron and carbon. Its properties mainly depend on extent of carbon. Along with carbon other elements such as Si, P, Mn, S and various transition elements etc. if present, wide varieties of steels with varying properties are formed.

2.5 NON-FERROUS ALLOYS

2.5.1 Copper Alloys

Physical properties and applications of copper alloys :

1. **Colour :** Red.
2. **Specific gravity :** 8.93 heavy metal.
3. **Melting point and boiling point :** M.P. = 1089°C, B.P. = 2350°C.
4. **Hardness :** Soft.
5. **Conductivity of heat and electricity :** Very good.
6. **Malleability and ductility :** Highly malleable and ductile.
7. **Any special property :** Tough and resistant to corrosion.
8. **Applications :** Alloys, electrical wires and cables, heating utensils, pans in making coins, ornaments, in electroplating and electrotyping etc.

2.5.2 Aluminium Alloys

Physical properties and applications of aluminium alloys :

1. **Colour :** Bluish white
2. **Specific gravity :** 2.7, light metal.
3. **Melting point and boiling point :** M.P. = 658°C, B.P. = 1800°C.
4. **Hardness :** Not very hard.
5. **Conductivity of heat and electricity :** Good.
6. **Malleability and ductility :** Malleable and ductile.
7. **Any special property :** Tough and high tensile strength, reflector of light.
8. **Applications :** Electric wires, cables for transmission lines, household utensils, air-craft, making alloys, Al foil for packing cigarettes and chocolates, in thermite etc.

2.5.3 Other Alloys

- Solders are low melting point alloys of tin and lead.
- These can be soft or hard depending upon the percentage of tin and lead.
- The solders become softer as the percentage of lead increases and that of tin decreases.
- Solder containing equal part of Pb and Sn is soft. Such solders have melting range between 180°-250°C and adhere to the metallic surface much better.
- An ordinary solder consists of Sn and lead in the ratio of 2 : 1. They are used for soldering electrical connections, sealing tin cans and joining lead pipes.

The most common solders are :

(i)   **Soft solders :** (Pb = 37-67%, Sn = 31.60%, Sb = 0.12%).

They melt at low temperatures. They are used for soldering electrical connections, sealing tin cans and joining lead pipes.

(ii)   **Brazing alloys :** (Sn = 92%, Pb = 5.5%, Cu = 2.5%).

Brazing alloys are used for soldering steel joints.

(iii) Tinmann's solder : (Sn = 66%, Pb = 34%).

It melts at 180°C and is used for joining articles of tin.

Low Melting Alloys :

1.   **Rose Metal :** (Bi = 50%, Pb = 28%, Sn = 22%).

It is a readily fusible alloy (m.p. = 89°C). It is used for making fire-alarms, fuse wires, castings for dental works and in automatic sprinkler's system.

2.   **Wood's metal :** (Bi = 50%, Pb = 25%, Sn = 12.5%, Cd = 12.5%).

Its melting point is low (71°C) and is easily fusible.

It is used (i) as a soft solder for joining two metallic parts, (ii) for making safety plugs of cookers, boilers, in fire-alarms and in water sprinklers etc., (iii) as casting in dental works, (iv) in electric fuse wires.

PRACTICE QUESTIONS

1.   Define metallurgy. Outline the general principles of metallurgy.
2.   (i)   Explain the terms : (a) Calcination (b) Roasting.
     What are their purposes ?
     (ii) With the help of figure, explain the process of calcination or roasting.
     (iii) Explain the stepwise process in roasting and calcination of an ore.
3.   Differentiate between :
     (a) Calcination and roasting. (b) Concentration and refining. (c) Gravity and electromagnetic separation.
4.   Write four physical properties of iron.
5.   **Short Answer Type Questions:**

     1.   Occurrence of iron
     2.   Extraction of cast iron
     3.   Blast furnace
     4.   Reactions in the blast furnace
     5.   Types of steels
     6.   Heat treatments on steel
     7.   Bessemer process
     8.   L.D. process
     9.   Commercial forms of iron
     10. Different varieties of iron
     11. Action of carbon monoxide on iron oxide
     12. The by-products of cast iron industry.

6.   **Long Answer Type Questions:**

     1.   Distinguish between annealing and normalizing.
     2.   Distinguish between Bessemer process and L.D. process.
     3.   Distinguish between steel and cast iron.
     4.   Explain the function of carbon monoxide in the manufacture of cast iron.
7.   Draw a neat sketch of a modern blast furnace and indicate, by means of equations, the chemical reactions taking place in its different zones.
8.   How does iron occur in nature?
9.   What are the sources of iron in India?
10. What is cast iron and what is its composition? What is the difference between white and grey cast iron?
11. In what ways Acid Bessemer process differ from Basic Bessemer process?
12. (a) Discuss the L.D. process for the manufacture of steel.
     (b) How are the impurities of pig iron removed in Acid Bessemer process?
13. (a) Steel-making processes are mainly divided into Acid and Basic process. Why?
     (b) Describe Basic Bessemer process for the manufacture of steel giving chemical reactions.
14. Describe a modern blast furnace for the manufacture of pig iron. Give a sketch of the furnace and label its various parts.
15. Describe fully the manufacture of cast iron from iron ore (red haematite), discussing the chemical reactions occurring in the blast furnace.
16. Point out the properties of steel which are modified by (i) Varying carbon content, (ii) Heat treatment.
17. Comment on the advantages of L.D. process. Sketch out the L.D converter and label it.
18. Before smelting iron, ore is roasted in free supply of air. Why?
19. L.D. process is far superior to Bessemer process. Why?
20. Iron does not occur in free state. Why?
21. Now-a-days steel is manufactured by L.D. process Why?
22. Alloy steels are called special steels. Why?
23. Steel articles are given suitable 'heat treatment'. Why?
24. Iron ores of India are regarded to be the best quality ores in the world. Why?

25. How does iron occur in nature? Discuss the manufacture of cast iron from haematite.
26. Define the term steel. How do you classify steels?
27. Discuss one important method of conversion of cast iron into steel.
28. Explain construction and working of blast furnace.
29. Discuss the chemical reactions involved in different zones of blast furnace.
30. Give the balanced chemical reactions involved in blast furnace during the manufacture of cast iron.
31. What is steel? Explain in brief types of steels.
32. Give the balanced chemical reactions involved in Bessemer process.
33. What are the products of blast furnace?
34. Draw neat labelled diagram of blast furnace or Bessemer furnace or L.D. converter. Mention advantages of L.D. process.
35. Give the important minerals of iron along with their chemical composition.
36. Explain the preliminary treatment involving the extraction of cast iron.
37. What is an alloy ? How alloys are classified ? Give two examples of alloys.
38. What are the purposes of making alloys ?
39. Give one example of an alloy having low melting point. State its three uses.
40. Define an alloy. Redefine it, if an alloy is considered as a solid solution.
41. Mention any four purposes of making alloys with at least one suitable example in each case.
42. Define alloy. Explain with example method for preparation of alloys.
43. Give the chemical composition of Wood's metal or Monel metal.
44. (a) Name two properties which are imparted to gold when alloyed by copper.
    (b) Write the important uses of (i) Babbit metal, (ii) Monel metal, (iii) Wood's metal, (iv) Duralumin.
45. Distinguish between an alloy and an amalgam. Give suitable examples where the casting property and tensile strength are increased by alloying.
46. Write the composition and applications of : (i) Monel metal, (ii) Wood's metal.
47. (a) Give purposes of alloying metals with suitable examples.
    (b) What are non-ferrous alloys ? Mention the principle constituents of Duralumin, Babbit metal and Phosphor Bronze.
48. Give the composition, properties and uses of the following non-ferrous alloys :
    (a) Duralumin, (b) Phosphor bronze, (c) Monel metal, (d) Wood's metal, (e) Babbit metal.
49. What do you understand by non-ferrous alloys ? Give the composition, properties and applications of Duralumin.
50. What is an alloy ? Describe the fusion method for the preparation of alloy.
51. Give the composition, properties and uses of Wood's metal.
52. Give the composition and four uses of Monel metal.
53. Give the composition, properties and uses of Babbit metal.
54. Give the composition, properties and uses of 'Alnico'.
55. Give the composition, properties and uses of (i) Duralumin, (ii) Dutch metal, (iii) German silver.
56. Write the composition, properties and uses of : (i) Gun metal, (ii) Babbit metal, (iii) German silver.
57. Give the properties and uses of Monel metal and Duralumin.
58. Give the composition of Wood's metal and Duralumin.
59. Write methods of preparation of alloys with one example.
60. Write any four applications of Duralumin, Wood's metal.
61. Classify the non-ferrous alloys and give one example of each.
62. State and explain the principle of compression method to prepare an alloy.
63. Explain the stepwise procedure for preparation of any alloy.
64. Define ferrous alloy giving suitable example. State the classification of alloys.

NON-METALLIC ENGINEERING MATERIALS

3.1 CERAMICS

3.1.1 Definition

- "Ceramics are inorganic non-metallic materials that are processed and used at high temperatures". In restricted sense, those articles, which are made of clay are mainly known as **'Ceramics'**.

- Even glass, cement and plastics could also be included in the ceramic materials.

3.1.2 Types of Ceramics

Ceramic materials and wares can be divided into the following basic groups according to their main fields of use.

1. **Structural ceramics :** Articles used mainly in constructing buildings and various other structures belong to this group e.g. building bricks, brick-blocks, hollow tile, roof tile, drain tiles etc. Rock goods, such as clinker brick, ceramic slabs for floors, sewer pipe etc.

2. **Facing material :** Articles used for internal and external facing of buildings and structures e.g. facing bricks and slabs and oven tiles.

3. **Refractories :** Materials which retain their mechanical properties at 1000°C or high temperatures. These are also used in making various parts of industrial furnaces, ovens and apparatus for operating at high temperatures.

4. **Fine ceramics :** Porcelain wares and glazed pottery are included in this class. They are used domestically (dishes, wash-basins, sinks, decorative articles), electrically (electrotechnical porcelain) and in laboratories (chemical ware and apparatus).

5. **Special ceramics :** A group of articles with specific properties utilised in radio industry, aviation instruments manufacture etc.

Subdivision of ceramics : For the sake of simplicity, ceramic materials may broadly be sub-divided into two classes.

1. **Heavy clay products :** Heavy clay products are those products which consist mainly of clay with small amount of other raw materials. Common bricks, roofing tiles, drain tiles, hollow tiles, sewer pipes, stone ware and refractories are well known examples of heavy clay products.

2. **Pottery products :** Those products which are made of such materials as terra cotta, earthen ware, porcelain, bone china and vitreous china are known as pottery products. The term pottery also includes wares made of earthenware, wall tiles, electrical insulators made of porcelain or other ceramic insulating materials. Sanitary ware, glazed wall and hearth tiles, electrical wares and refractory goods are well known examples of pottery products.

In general, pottery may be classified into the following four groups.

(a) **Terra cotta :** Terra cotta includes all potery wares made form common clays. Terra cotta comparises all porous pottery ware, which are not covered with a glaze. It is not subjected to higher temperatures to allow the body of the ware to become impermeable to liquids. It can be scratched by a hard steel. Its colour varies from light yellow to red, reddish brown and brown. Common bricks, tiles and hollow ware made by **kumhars** belong to this class.

(b) **Earthen ware :** All properties that are permeable or porous, made from red burning clays and white clays coated with a glaze are included in earthen ware. It is harder than terra cotta, because it is fired at higher temperature.

(c) **Stone ware :** Potteries which become impermeable to liquids during firing are called stone wares. It is greyish or brownish body. They are dense and often glazed with salt glaze. Drain water pipes, zars, carboys for keeping liquids, specially acids, sanitary wares, such as wash basins etc. are the examples of stone ware.

(d) Porcelain : Porcelain is a pottery with a white, translucent and impermeable body. They are the best type of potteries and the composition of the body is so adjusted that when subjected to high temperature it becomes completely translucent to light.

3.1.3 General Properties of Ceramics

1. **Chemical and Physical properties :** The components present in ceramics, such as oxides, carbides etc. give high chemical stability to ceramics. Most of the constituent oxides are usually resistant to highly oxidising and reducing atmospheres and also to fluctuations in temperatures. Compactness of the crystal structure, high directional character of chemical bonding and high field strengths of small cations of high charge are also responsible for the stability of ceramics.

2. **Mechanical properties :** Ceramics are brittle solids which are very resistant to compression. The strength of ceramic is mainly controlled by the following important factors :

   (i) temperature, (ii) size and shape, (iii) composition, (iv) surface conditions and (v) microstructure.

3. **Electrical properties :** Oxide ceramics are generally bad conductors or insulators in their normal oxidation states. The non-oxide ceramics, act as semiconductors.

3.1.4 Basic Raw Materials Used in Ceramics

The three main raw materials used in making the ceramic products are clay, feldspar and sand.

1. **Clay :** Clay is a material of great importance and is used mainly in the production of porcelain, stone ware and earthen ware. From a ceramic point of view, clays are plastic and mouldable when sufficiently finely pulverised and wet, rigid when dry, and vitreous when fired at a higher temperature.

On the basis of their structure, clays are classified into the following three major types of minerals :

(i) Kaolinite : $Al_2O_3, 2SiO_2 \cdot 2H_2O$.

(ii) Montmorillinite : $(Mg, Ca) O \cdot Al_2O_3 \cdot 5\ SiO_2 \cdot nH_2O$.

(iii) Illite : $K_2O \cdot MgO \cdot SiO_2 \cdot H_2O$.

Obviously, clays from different sources have different properties. By virtue of their physical and refractory properties, the following types of clay are usually distinguished.

(a) Kaolin or china clay   (b) Pottery clay

(c) Fuller's earth   (d) Bentonite

(e) Fire clay   (f) Ball clay

(g) Lithomarge.

From the point of view of ceramic industry, kaolins, pottery clays and ball clays are most important.

2. **Feldspars :** These are the main constituents of igneous rocks and are abundent and wide spread in the earth's crust. Feldspars do not occur in pure state in nature. The following are three major types of feldspars, which are used in ceramic industry :

(1) Potash feldspar : $K_2O \cdot Al_2O_3 \cdot 6\ SiO_2$.

(2) Sodium feldspar : $Na_2O \cdot Al_2O_3 \cdot 6\ SiO_2$.

(3) Lime feldspar : $CaO \cdot Al_2O_3 \cdot 2\ SiO_2$.

Feldspar is used in ceramic bodies in lesser proportions than silica. This is extensively used in white wares.

3. **Silica (SiO_2) :** It is the third main constituent of ceramic material. Silica occurs abundently in the form of quartz (rock crystal) in volcanic rocks and consequently also in clays. Silica is used in the ceramic industry in the form of sand, sandstone, quartz and flint.

PORCELAIN :

Composition of porcelain :

Kaolin = 40 %, Feldspar = 30 %, Flint = 20 % and Ball clay = 10 %.

No additional refractory material is required because kaolin itself is a refractory. In the compounding of body, the amount of kaolin, quartz and feldspar should be adjusted according to their actual composition.

Porcelain is made from the purest white clay agglutinised by some substances such as powdered feldspar, which softens and fuses at the temperature at which the ware is fired, vendering the mass semi-transparent. Glazing improves the appearance, surface hardness and electrical properties of porcelain.

The pure porcelain is called mullite or sillimanite. Excellent types applied to electrical uses are isolantite and alsimag.

Properties (of sillimanite) :

1. High softening point i.e., above 1800°C.
2. Perfectly stable up to softening point.
3. Coefficient of expansion 0.45×10^{-5} and is able to withstand sudden changes in temperature.
4. It has got neutral reaction.
5. Highly resistant to corrosive action of many slags and is resistant to abrasion.
6. Stable in oxidising and reducing atmospheres. Resistant to the action of chemicals.
7. Good strength upto softening point.
8. Very low electrical conductivity, hence a good electrical insulator.
9. High thermal conductivity.
10. Pure sillimanite is colourless but generally brown in colour due to the presence of iron in it.

Uses :

1. Because of its refractory properties it is used in making refractory blocks, bricks, crucibles, saggers and other refractory fittings.
2. Because of its electrical insulation property it is used for making suspension insulators, pin type insulators, transformer bushings and spark plug insulation.
3. Because of its resistance to chemical action it is also used for making jars and components for chemical reactions.
4. Porcelain is also used for many dental applications.

3.2 REFRACTORIES

3.2.1 Definition

- "In the broad sense, refractory is any material that can withstand high temperatures without softening or suffering a deformation in shape."

3.2.2 Characteristics

- A good refractory possesses the following characteristics :
1. It should be infusible at the temperature to which it is to be exposed.
2. It should not crack and suffer loss in size at the operating temperatures, because such defects might lead to the destruction of a complete furnace, arch or roof.
3. It should resist the abrading action of the gases, flames, molten metals, slags etc.
4. It should be chemically inert towards corrosive action of gases, molten metals and slags.
5. It should be able to withstand the load of structure or pressure at operating temperatures.
6. The expansion and contraction of the refractory should not be sudden, but it should be uniform with rise and fall of temperature.

3.2.3 Applications of Refractories

- In metallurgy and for various industries, refractories are the essential materials of construction. Their main objective is to resist loss of heat and at the same time to resist the corrosive and abrasive action of molten metals, slags and gases at high operating temperatures without undergoing softening or distortion in shape.

- Refractories are mostly used for the construction of the lining of the furnaces, tanks, boilers, converters, pyrometer tubes, kilns, crucibles, ladles, etc. used for the manufacture of metals, cement, lime, glass, ceramics, paper etc.

3.2.4 Classification of Refractories

- Depending upon the chemical nature of the main constituents (mostly oxides and their compounds), the refractory materials are classified under three general groups :

    (1)   Acid refractories

    (2)   Basic refractories.

    (3)   Neutral refractories.

    (1)   Acid refractories : are those which consist of acidic materials like alumina (Al_2O_3) and silica (SiO_2). They are not attacked by acidic materials, but are easily attacked by basic materials. Silica is the principal constituent of acid refractories.

    Examples : Fire-clays, silica, alumina, quartz, gainster etc.

    (2)   Basic refractories : are those which consist of basic materials like CaO, MgO. These are not attacked by basic materials, but are easily attacked by acidic materials.

    Examples : Bauxite, magnesite, dolomite etc.

    (3)   Neutral refractories : are those which consist of substances which do not combine either with acidic or basic oxides. Neutral refractories are made from weakly acidic or basic material like carbon.

    Examples : Graphite, chromite, zirconia, carborundum (SiC) etc.

3.2.5 Properties of Refractories

- The important characteristic properties of refractories are usually taken into consideration according to the use of which they are applied. These are given below :

    (1)   Refractoriness : "It is the ability of the material to withstand the action of heat without appreciable deformation or softening under particular service conditions." The softening temperature or melting point of the material is generally considered as the measure of its refractoriness.

    Most of the common refractory materials are mixtures of several metallic oxides, hence they do not have a sharp fusion temperature. It is common practice to determine softening temperature rather than fusion temperature. The softening temperatures of refractory materials are, generally, determined by using pyrometric cones test. It is necessary that a material to be used as refractory should have softening temperature much higher than the operating temperature of the furnace in which it is to be used. It is however, noteworthy that in the inner refractory lining in a furnace is at a much higher temperature than the outer ones. Therefore, unless the refractory material melts away completely, it can often be used to withstand a temperature higher than its softening temperature. Since the outer end of refractory is at a lower temperature and still in solid state provides required strength.

    (2)   Strength or refractories under load : The refractories are mostly subjected to bear loads, tension and shearing stresses particularly at high temperatures. So, it is essential that refractory materials should be strong enough to bear such loads without any deformation and have a good resistance to the tension and shearing stresses which may be imposed during the operation.

    (3)   Thermal expansion and contraction : All solids generally expand when heated and contract when cooled, so in furnace design, allowance has to be made for thermal expansion. The expansion effects all dimensions (i.e. length, area and volume) of a body. Therefore, it is necessary that a refractory material should have least possible thermal expansion. Fire-clay, alumina and magnesite refractories tend to shrink (contract) while silica brick may expand when heated. The expansion and contraction may result in damaging the joints and producing cracks so that the whole structure of the furnace may be spoiled. So a good refractory should have low expansion and contraction.

(4) Thermal conductivity : Thermal conductivity of a refractory determines the amount of heat that will flow through a furnace wall under given conditions and the knowledge of this property is essential for designing the furnace.

The refractories with low thermal conductivity will not allow much loss of heat of the furnace. So the refractories of low thermal conductivity are mostly used for the furnaces such as blast furnace, open hearth furnace etc. But for muffle furnace walls, retorts, coke-oven walls, the refractory of high thermal conductivity is required, which will transfer heat form the outer surface to the charge. To prevent loss of heat, sometimes a refractory is backed by an insulating material like asbestos.

(5) Porosity : Porosity of a refractory material is the ratio of its pores volume to the bulk volume. It is the important property of refractory bricks because it affects many other properties e.g. chemical stability, strength, abrasion resistance and thermal conductivity. In a porous refractory, molten charge, slag, gases etc. are likely to enter more easily and to a greater depth and may react and reduce the life of refractory material. Porosity decreases the strength, resistance to corrosion or penetration by slags, gases, etc. but increase resistance to thermal spalling. (i.e. thermal shock resistance). Moreover, the least porous bricks have the highest thermal conductivity, due to the absence of air voids. In porous bricks the entrapped air in the pores, acts as a non-conducting material of heat. A good refractory, in general, should have lower porosity.

(6) Thermal spalling : It is the cracking, breaking, peeling off or fracturing off a refractory brick or block under high temperature. spalling is, generally, due to rapid changes in temperature, which causes uneven expansion and contraction within the mass of refractory material, thereby internal stresses and strains are developed in it. Spalling may also be due to the penetration of slag into refractory brick,-thereby causing variation of coefficient of expansion. So a good refractory must show a good resistance to thermal spalling.

(7) Chemical inertness : The refractory material selected for a particular purpose should be chemically inactive in use and does not easily form fusible products with slags, fuel ashes, furnace gases etc. and corrode it.

(8) Resistance to abrasion or erosion : Refractory to last longer, it is desirable that it is least abraded by descending hard charge, flue gases escaping at high speeds, particles of carbon or grit etc. Resistance to abrasion (or erosion) is very important for such constructions as by-product coke-oven walls and linings of discharge end of rotary kiln etc.

(9) Electrical conductivity : The refractory material to be used for lining electric furnaces should have low electrical conductivity. Except graphite, all other refractories are poor conductors of electricity However, electrical resistance of refractories decreases rapidly with rise of temperature.

3.2.6 Uses

(1) Fire-clay bricks : These are made from finely ground soft plastic material fire clay ($Al_2O_3 \cdot 2SiO_2 \cdot 2H_2O$) with powdered calcined fire-clay called 'grog'. The exact proportions of the constituents depend on the type of bricks to be made. Greater is the percentage of grog, the lesser will be the spalling tendency. General composition of fire-clay bricks ranges from 55% SiO_2 and 35% Al_2O_3 (feebly acidic) to 55% Al_2O_3 and 40% SiO_2 (nearly neutral bricks), the balance consists of necessary oxides in the clay like $K_2O \cdot FeO$, CaO, MgO etc.

Properties : Fire-clay bricks are light yellow to reddish brown in colour, depending on the contents of iron oxides. They are slightly acidic in character because of SiO_2 content. They possesses low porosity and lower refractoriness than silica bricks. They usually fuses at 1350°C under a load of 2 kg/cm^2. Their crushing strength is quite high (about 200 kg/cm^2). A very useful property of fire-clay bricks is better resistance to thermal spalling than silica bricks and this makes them very suitable for making a checker works of regenerative furnaces and charging doors etc. which are subjected to temperature fluctuations. Fire-clay bricks are much cheaper than silica bricks.

Uses : Fire-clay bricks are most widely used of all the refractory bricks, since they are very cheap and are well suited for many applications. They are used for construction of blast furnaces, open-hearth furnaces, stoves, ovens, crucible furnaces, flues, kilns, boiler-settings, regenerators, charging doors etc.

(2) Silica bricks : Contains 90 to 95% SiO_2 and about 2% lime is added during grinding to furnish the bond. Basic materials used for their manufacture are quartz, quartzite, ganister, sand, sand stone etc. For their manufacture the siliceous rocks is crushed and ground with 2% lime and water. The thick paste is then made into bricks either by hand moulding or by machine pressing. The bricks are dried in air and then burnt in kilns. During heating, temperature is slowly raised, in about 24 hours, to about 1500°C, and this is maintained for nearly 12 hours, so as to allow quartzite to be converted into cristobalite. Then cooling is done carefully and it takes about 1 to 2 weeks. During cooling, cristobalite is slowly changed into tridymite, so that a mixture of tridymite and cristobalite results in the final bricks.

Properties : Silica bricks are yellowish in colour with brown speck throughout the body and contains about 25% pores. Silica bricks do not contract in use, but they have permanent expansion of about 15% when reheated. This effect is reversible and the brick returns to its original size when cooled. This expansion is caused by reversible allotropic transformations.

$$\text{Quartzite} \underset{}{\overset{87°C}{\rightleftharpoons}} \text{Tridymite} \underset{}{\overset{117°C}{\rightleftharpoons}} \text{Cristobalite}$$

$$\text{(crystalline)} \qquad \text{(}\alpha\text{-form of silica)} \qquad \text{(}\beta\text{-form of silica)}$$

$$\text{sp. gr} = 2.65 \qquad \text{sp. gr.} = 2.26 \qquad \text{sp. gr.} = 2.32$$

Thus, if during firing of silica bricks quartzite is not converted into tridymite and cristobalite, the bricks will expand to the extent of about 17.2% during use in the furnace and consequently, the refractory structure will fall. That is why heating at 1500°C for about 12 hours is necessary, during the making of silica bricks.

Silica bricks have homogeneous texture, free from air-pockets and moulding defects. They can withstand a load of about 3.5 kg/cm^2 upto about 1500 - 1600°C. Thus, they are remarkable for their load bearing capacity especially at high temperatures. Silica bricks are relatively light (sp. gr. = 2.3 to 2.4) and possess high rigidity and mechanical strength.

Uses : The main applications of silica bricks are roofs of open-hearth furnaces, open-hearth steel, making furnaces, coke-oven walls, cowper stoves, domes, roofs of electric furnaces, linings of acid convertors, glass furnaces etc.

(3)   Masonry bricks : There are various types of bricks used in masonry :

1.   Common burnt clay bricks.

2.   Sand lime bricks (calcium silicate bricks).

3.   Engineering bricks.

4.   Concrete bricks.

5.   Fly-ash clay bricks.

Bricks are used as siding in the building industry due in part to its important characteristics and just because it can be a good affordable option. Below we summarize the benefits and applications of the most commonly used type of bricks.

1.   Common burnt clay bricks :

2.   Sand lime bricks :

Sand lime bricks are made by mixing sand, fly ash and lime followed by a chemical process during wet mixing. The mix is then molded under pressure forming the brick. These bricks can offer advantages over clay bricks such as :

- Their colour appearance is gray instead of the regular reddish colour.

- Their shape is uniform and presents a smoother finish that does not require plastering.

- These bricks offer excellent strength as a load-bearing member.

3.   What are engineering bricks ?

Engineering bricks are bricks manufactured at extremely high temperatures, forming a dense and strong brick, allowing the brick to limit strength and water absorption.

Engineering bricks offer excellent load bearing capacity, damp-proof characteristics and chemical resisting properties. These bricks are used in specific projects and they can cost more than regular or traditional bricks.

4.   Concrete bricks :

Concrete bricks are made from solid concrete and are very common among homebuilders. Concrete bricks are usually placed in facades, fences and provide an excellent aesthetic presence. These bricks can be manufactured to provide different colours as pigmented during its production.

5.   Fly-ash clay bricks :

Fly-ash clay bricks are manufactured with clay and fly ash, at about 1,000°C. Some studies have shown that these bricks tend to fail poor produce pop-outs, when bricks come into contact with moisture and water, causing the bricks to expand.

3.3 COMPOSITE MATERIALS

3.3.1 Definition

- Each class of basic engineering materials (metals, high polymers and ceramics) has its own outstanding and distinct characteristics as well as limitations; and any of these cannot be properly used. Where very strangent and specific requirements are needed e.g., gas turbines, high temperature reactors, supersonic aircrafts, missles, space-crafts, etc. In order to meet such very *strangent* and *specific* requirements, technologists and scientists have developed a new class of materials, called "composites".

- **A composite material** may be defined as a material system consisting of a mixture of two or more micro-constituents, which are mutually insoluble, differing in form and/or composition, and forming distinct phases.

3.3.2 Properties

- Composite material usually, possesses properties different in kind and scale from those of any of its constituents. Composite materials possess such combination of properties, which were not within the easy reach of engineers either because such basic materials were not be available or were too costly. In composites, stiffness (the ability of a material to resist elastic deformation or deflection on loading) is increased, without the disadvantages of brittleness. Thus, by using composites, it is possible to have such combination of properties like high-strength and stiffness, even at elevated temperatures : corrosion-resistance and ability to withstand extreme-temperature conditions; desirable coefficient of thermal expansion, etc.

- Wood (a composite of cellulose fibres and lignin cementing materials), bone (a composite of a soft, but strong protein callogen, and brittle and hard mineral apatite), packing paper impregnated with bitumen or ware rainproof cloth (cloth impregnated with water-proof material), insulating tape, wase concrete, etc. are some of the common examples of composite materials.

3.3.3 Types of Composites

Following three types of composites are, generally, made :

(i) Particulate-composites are made by dispersing particles of varying size and shape of one materials in a matrix of another material. The operation may be carried out by adding particles to a liquid matrix material, which later solidifies (during age-hardening) or may be pressed together (by the powder process). In such composites, the matrix as well as particles share the load-bearing function. Examples of such composites are tungsten-thoria (used as lamp filaments), cermets (metal-ceramic composites) used in cutting tools, concrete used in dams, building, etc. The effective strength of such composites depends to a great extent on interparticle spacings.

As the volume fraction of the dispersoids increases, the mechanical properties improve, reaching an optimum value and then begin to fall. At higher volume fractions of the dispersoids, the brittle ceramic particles come closer to each other, thereby somewhat continuous brittle phase is formed, which promotes premature failure. For example, in tungsten carbide-cobalt (WC-Co) composite cements, the maximum strength is attained around 66% volume percentage of WC, after which it begins to fall.

(ii) Layered-composites are extensively employed in many familiar applications. For example, plywood (a laminated composite of thin layers of food with alternate layers, having different orientation of the grains), stainless steel cooling vessel with a copper clad bottom, and stainless steel bonding on both sides of copper, etc.

(iii) Fibre-reinforced-composites have three components, namely, filament, a polymer matrix (encapsulating agent for the filaments), and a bonding agent (which ties fibre filaments to polymer). Glass fibres and metallic fibres are commonly employeds for this purpose. The fibres can be employed either in the form of continuous lengths, staples or whiskers.

Fibres in composites provide stiffness and also lowers the overall density of the composites. Where metal-matrix is used. 'Matrix' provides toughness to the composite material and it binds the fibres together. Thermosetting and thermoplastics like polyster resin, epoxy resin. phenolics silicons etc. are commonly used as matrix materials. The most familiar example of this category of composites is garrs-reinforced plastics aluminium born composites. Boron has low density and is suitable for light-weight applications. Its elastic modulus is one of the highest of the elements, but it is brittle. When it is used as a reinforcing fibre for aluminium (a ductile matrix) the elastic modulus of composite is increased. At the same time the disadvantages of brittleness of boron are encountered by the cushioning effects of the ductile matrix of Al. The ductile matrix of Al stops any propagation of cracks, if a fibre of B embedded in it, breaks accidentally. If the entire material were of boron only a propagating crack would cause ultimate failure of the entire cross-section.

The fibre-rienforced composites possess superior properties like higher yield strength fracture strength and fatigue life. The fibres prevent slip and crack propagation and inhibit it, thereby increasing mechanical properties. When a load is applied, there is localized plastic flow in matrix, which transfers the load to the fibres embedded in it. When a soft phase is present in hard matrix, the shock-resistance of the composite is increased. On the other hand, if hard reinforcing fibres are present in a soft matrix, the strength and modulus of composites are increased. To obtain composite having the maximum strength and modulus, it is essential that there should be maximum number of fibres per unit volume, so that each fibre takes the full share of the load. The fibre-reinforced composites are, generally anisotropic (i.e. having different properties in different directions) and the maximum strength is in the direction of alignment of fibres. For getting isotropic properties, the fibres are oriented randomly within the matrix, e.g. in ordinary fibre glass. It may be pointed here that the cost of laying fibres aligned in a particular direction is much higher than that for random orientation. For preparing a fibre reinforced composite. It is essential that : (i) the coefficient of expansion of the fibre matches closely that of the matrix (ii) the fibre and matrix would be chemically compatable with each other and no undesirable reaction takes place between them, (iii) the fibre should be stable at room temperature and retain a good percentage of strength at elevated temperatures.

3.3.4 Advantageous Characteristics of Composites

- Important advantages of composite materials over the bulk materials (metals, polymers and ceramics) are : (1) higher specific strength, (2) lower specific gravity, (3) higher specific stiffness, (4) maintain strength, even upto high temperatures, (5) better toughness impact and thermal shock-resistances, (6) cheaply and easily fabricable, (7) better creep and fatigue strength, (8) lower electrical conductivity, (9) lower thermal expansion, (10) better corrosion and oxidation resistance.

3.4 ADHESIVES

3.4.1 Definition

(Nov. 18)

- "Adhesive is a substance used for sticking two unlike bodies together, due to the molecular forces existing in the area of contact".

OR

- An adhesive can be defined as "any substance capable of holding material together by surface attachment". The bodies held together by an adhesive are known as **'adherends'**, while the process of holding one adherent to another by adhesive, is called **'bonding'** and the final assembly of two adherends and adhesive is called **'bond'** and sometimes 'joint' also.

3.4.2 Properties

- The quality of adhesive is judged by (a) degree of tackiness (i.e. stickness), (b) rapidity of bonding, (c) strength of bond setting on drying and (d) durability.

3.4.3 Advantages

(1) Adhesives have advantage over other joining methods in that they can be applied to the surfaces of any material and such materials may be joined together as glass and metal, metal and metal, metal and plastic, plastic and plastic, ceramic and ceramic.

(2) The structural members joined by adhesives are notably free from any residual stresses. This makes it possible to fully utilise the adherent strength of any material.

(3) In several cases of bonding by adhesive, no high heat is required.

(4) Surfaces are easily and rapidly attached to each other by adhesive.

(5) Adhesives introduce heat as well as electrical insulating layers in between the bonding surfaces.

(6) The process of applying adhesives is very simple and does not require a number of highly specialized persons.

(7) Adhesive bonding requires less time after finishing as compared with other joining processes like welding, soldering etc.

(8)   Bonding by adhesive gives smoother surface. So bonding by adhesives are better suited in articles such as brake and clutch facings or aircraft assemblies, where rivet heads or other projections affect seriously their performance.

(9)   New and better composite articles can be designed by efficient bonding together various materials by adhesives to get the advantage of best properties of each material e.g. flush-doors, metal-faced plywoods etc.

(10)   By bonding together dissimilar metals with adhesives, galvanic corrosion of dissimilar joints is reduced by preventing direct metal to metal contact.

(11)   Adhesive joints are leak-proof for gases and liquids. So adhesive bonding of components is used in preparing water-tight wooden boats etc.

(12)   Metal joined by adhesive can resist corrosion.

3.4.4 Limitations of Adhesive Bonding

(1)   Because most of the adhesives are organic materials, so they cannot be used at high temperatures and their bond strength decreases rapidly as the temperature rises.

(2)   There is no single general purpose adhesive, which can join all types of surfaces. So for a particular job particular suitable adhesive is required and this may also require particular bonding conditions. Hence, careful selection of an adhesive, for each set of substances to be bonded is necessary.

(3)   Adhesives often require appreciable time for their reactions of setting, curing etc. and hence they do not develop their full bonding strength and performance, immediately after application, unlike welding or mechanical joining by rivets etc.

(4)   Adhesives are, generally, susceptible to high humidity.

(5)   Only plain and very clean surface can be joined for a reliable strength by the use of adhesives.

(6)   Adhesive strength is generally lower than other joining methods like welding, rivetting etc.

(7)   Testing and inspection of joints, bonded by adhesives, is rather difficult.

3.4.5 Examples of Adhesives

1.   Phenol-formaldehyde resin : It is sold as a solid, liquid or impregnated film. The resin is first coated on the two surfaces to be bonded and then, cured by heating and pressing. The bond film of resin formed is hard, highly resistant to the action of insects, fungi, water etc.

Uses : For making water-proof plywoods, laminates and bonding articles in aircraft and ship building industries.

2.   Urea-formaldehyde resin : It is a transparent, syrupy compound, which can be used as such or after mixing with water. The addition of water helps in the formation of cross-linkages promotes rapid setting and also helps in avoiding cracks in the bond film produced. It can be used in cold but a little heating helps in accelerating the setting process. Urea formaldehyde on condensation with $ZnCl_2$ yields a solid adhesive. The bond film produced by urea-formaldehyde resin is (i) quite strong, (ii) good resistant to moisture, insects and fungi. However, action of acids or alkalies deteriorate the resin film after some time.

Uses : For bonding wooden surfaces, plywoods, laminates, articles of aircraft and ship industries etc.

3.   Epoxy resins : Epoxy resins have good adhesive properties. They (i) have ability of getting cured, without the application of external heat, (ii) have good resistance to chemicals, (iii) have low shrinkage during curing and (iv) possess good electrical resistance. They stick to a number of substances including metals and glasses. Epoxy compounds may be used in solid or liquid form. Usually, they are modified by adding some external compounds (like unsaturated fatty acid or amine) and some solvents. The solvent evaporates, leaving behind a very thin film, which possess excellent adhesion.

3.4.6 Uses

For bonding glass, metallic and ceramic articles. A trade name for a common epoxy resin type adhesive is **'araldite'**, which is much used in aircraft industry.

PRACTICE QUESTIONS

1. What do you understand by ceramics ?

2. What are the basic raw materials used in ceramics ? Write their important properties.

3. What is the chemical composition of porcelain ? Give its properties and uses.

4. Explain the essential requirements of good refractory.

5. Give main applications of silica bricks.

6. Silica refractories have low resistance to spalling. Give reasons.

7. How do porosity, spalling, strength and thermal conductivity influence the nature and utility of refractory ?

8. Describe the manufacture, properties and uses of fire-clay refractories.

9. Define composite materials. What are the advantageous characteristics of composites ?

10. What are adhesives ?

11. Mention the advantages and limitations of adhesives.

12. Give the properties and applications of :

   (i)   Phenol-formaldehyde resin

   (ii) Urea-formaldehyde resin and

   (iii) Epoxy resins.

13. Mention the uses of epoxy resins.

QUESTIONS WITH ANSWERS

1.   What is an adhesive ?

Ans. Any substance capable of holding materials together by surface attachment.

2.   Define the terms : adherends, bonding and bond.

Ans. The bodies held together by an adhesive are known as 'adherends', while process of holding one adherend to another by adhesive is called 'bonding' and the assembly of two adherends and adhesive in between them is called a 'bond'.

3.   Name any two adhesives used for bonding wooden surfaces, plywoods and laminates etc.

Ans. Phenol-formaldehyde resin and urea-formaldehyde resin.

4.   How phenol-formaldehyde is used as adhesive ?

Ans. The resin is first coated on the two surfaces to be bonded and then cured (cross-linked) by heating and pressing.

5.   What type of adhesive is used for bonding glass, metals and ceramics ? Give an example.

Ans. Epoxy resin e.g. araldite.

6.   Mention one drawback of urea-formaldehyde resin over phenol-formaldehyde resin as an adhesive.

Ans. Urea-formaldehyde resin film deteriorates after some time by the action of acids or alkalies.

7.   Which of the following adhesives possess maximum bond strength ? Phenol-formaldehyde resin, urea-formaldehyde resin, epoxy resin.

Ans. Epoxy resin.

WATER

4.1 INTRODUCTION

- Water is the most useful and also the most abundant of all chemical substances. It plays an important part in a wide variety of natural processes and is essential for the growth of all animal and plant life.
- Water present on earth passes through a remarkable cycle of changes. Thus from sea water it forms water vapour, then forms a cloud, from where it comes down as rain water and flows as mineral water, river water and finally comes again as sea water.

4.2 SOURCES OF WATER

- Water on the earth is classified as (1) Surface water, (2) Underground water. The surface water includes rain water, river water, lake water, sea water; while underground water is mainly well water and spring water.

1. **Surface water :**

   (a) Rain water : This is the purest form of water. It is obtained by natural process of evaporation and condensation water. When it reaches earth from atmosphere, it dissolves gases like CO_2, SO_2, NO_2, etc. from the atmosphere and carries suspended solid particles with it.

   (b) River water : River gathers water mostly from rain and some part from spring water. When water flows over earth's surface, it dissolves soluble minerals from soil. The greater the contact of soil containing soluble salts, greater will be the amount of dissolved salts in river water. River water thus contains dissolved salts like chlorides, magnesium, iron, etc. It carries with suspended soil particles and organic matter derived from decomposition of plant and animal bodies.

   (c) Lake water : This is the water stored above earth's surface. It contains lower quantities of dissolved salts or minerals than well water, but quantity of organic matter suspended is quite high. The lake may collect some spring water also. Lake water can be directly used for domestic or industrial purpose.

   (d) Sea water : This is the most impure form of natural water. Rivers throw all the impurities carried with water into sea. The continuous evaporation of water from surface of sea increases concentration of dissolved impurities. This sea water contains about 3.5% of dissolved impurities, out of which about 2.5% is sodium chloride. Thus sea water do not be used directly for domestic or industrial purpose.

2. **Underground water :**

- A part of rainwater, which reaches earth's surface or water from rivers percolates into the earth. This water travels downwards, during the flow of it comes in contact with various mineral salts present in the soil and dissolves some of them. Water continues its downward flow till it meets hard rock. When it retreads upward and may come out in the form of spring.
- Spring and well water is generally clear in appearance as it is filtered through soil layers, but contains considerable quantity of dissolved salts. The water from well and spring contains more hardness. The underground water is suitable for domestic use. Some spring water contains colloidal sulphur and shows medicinal value.

4.3 IMPURITIES OF WATER

- The common impurities present in the natural water may be classified as follows : (1) Suspended impurities, (2) Dissolved impurities, (3) Colloidal impurities, (4) Biological impurities.

1. **Suspended impurities :**

- Water may contain the suspended impurities such as clay, mud, algae, industrial waste, organic matter etc. These remain suspended in water. They produce turbidity, colour and odour and may cause diseases.

2. **Dissolved impurities :**

   - There may be two types of dissolved impurities viz. (a) Gases, (b) Mineral salts.

   (a) Gases : The gases present in air such as oxygen, carbon dioxide, oxides of nitrogen, hydrogen sulphide, sulphur dioxide, are soluble in natural water.

   (b) Mineral salts : The commonly observed salts in natural water are carbonates, bicarbonates, sulphates and chlorides of calcium, magnesium, sodium and potassium. When salts of calcium and magnesium are present in water, it is known as hard water. Such a water cannot be used for washing clothes and in boilers.

3. **Colloidal impurities :**

   - Colloidal particles of clay, fine mud, decayed leaves, organic matter (insects and fungi) may be present in water.
   - Such impurities neither settle down on standing nor can these be removed by filtration.
   - Size of colloidal particle is about 10^{-4} cm to 10^{-7} cm.
   - These are the charged particles. These become the main source of epidemic when associated with bacteria.

4. **Biological impurities :**

   - Slime growth such as algae, micro-organism i.e. bacteria, fungi etc. are considered as pathogenic.
   - Such impurities may be present in natural water and it may affect the efficiency of heat transfer equipment.
   - Biological growth can also contribute to corrosion and cause odours of water which might restrict its usefulness.
   - Such water is dangerous for human consumption and health.

4.4 HARD AND SOFT WATER

(Nov. 18)

1. **Soft water :** Water which does not contain any of the calcium or magnesium salt dissolved in it is called as soft water. Soft water may contain salt like sodium or potassium dissolved in it. These salts have no effect on soap solution.

2. **Hard water :** Water containing the dissolved salts or calcium and magnesium is called as hard water. Due to presence of these calcium and magnesium salts, hard water cannot produce good lather or foam with soap. Hardness in water is that characteristic which prevents the lathering of soap. A soap is generally a sodium salt of long chain fatty acids such oleic acid, palmetic acid and stearic acid. The soap consuming capacity of water is mainly due to reaction of hardness causing calcium and magnesium ions with soap i.e. sodium salt of long chain fatty acids. They form insoluble salts of calcium and magnesium soaps which do not have detergent value.

   The reaction of consumption of soap by calcium chloride or magnesium sulphate can be given as follows :

   $$2\ C_{17}H_{35}COONa + CaCl_2 \ \rightarrow \ [C_{17}H_{35}COO]_2\,Ca + 2\ NaCl$$

   Soap   Impurity   Insoluble salt   Calcium

   Sodium stearate in water   stearate

   $$2\ C_{17}H_{35}COONa + MgSO_4 \ \rightarrow \ [C_{17}H_{35}COO]_2Mg + Na_2SO_4$$

   Soap   Hardness   Magnesium stearate

   The water which does not produce lather with soap solution readily, but forms white curd is called hard water.

4.5 HARDNESS OF WATER

(Nov. 18)

- It is a common observation that water from some sources like wells produce lather (foam) readily with soap while that from other sources like rivers, springs or even certain wells produces lather with soap only with difficulty.

- Water which does not produce good *lather* with soap solution *readily* but develops *curd* (a white scum) is termed as *hard water*. Water which does not produce curd with soap solution but develops lather *readily* is known as *soft water*.

4.6 CAUSES OF HARDNESS OF WATER

- Rain water absorbs carbon dioxide from air and also from decaying plants on soil. Such water, when flows over the rocks containing calcium and magnesium carbonates, reacts slowly with these substances forming bicarbonates which are much more soluble.

$$H_2O + CO_2 \longrightarrow H_2CO_3$$

$$H_2CO_3 + CaCO_3 \longrightarrow Ca(HCO_3)_2$$

(Soluble)

- On the surface layer, there are also chlorides and sulphates of calcium and magnesium and these salts are soluble in water.
- Thus, water flowing over rocks and surface layer contains bicarbonates, chlorides and sulphates of calcium and magnesium in the dissolved state. Such water is known as *hard water*, which does not produce good lather with soap solution.
- Soap is a mixture of sodium salts of fatty acids like stearic acid, oleic acid, palmitic acid etc. Soap (sodium salt) is soluble in water and forms lather due to which it has cleansing property.
- When soap is added to hard water containing calcium and magnesium salts (i.e. Ca^{++} and Mg^{++}), insoluble calcium and magnesium salts are obtained by double decomposition.

$$2C_{17}H_{35}COONa + CaCl_2 \rightarrow Ca(C_{17}H_{35}COO)_2 \downarrow + 2NaCl$$
(Soluble sodium stearate; i.e. soap)   (Insoluble calcium stearate)

$$2C_{17}H_{35}COONa + Ca(HCO_3)_2 \rightarrow Ca(C_{17}H_{35}COO)_2 \downarrow + 2NaHCO_3$$
(Insoluble calcium stearate)

$$2C_{17}H_{35}COONa + MgSO_4 \rightarrow Mg(C_{17}H_{35}COO)_2 \downarrow + Na_2SO_4$$
(Insoluble magnesium stearate)

- This insoluble calcium or magnesium stearate is known as curd i.e. white scum. Until all the calcium or magnesium ions are removed in this way, soap will not be available for cleansing purposes. Thus, much soap is wasted in hard water.
- Moreover clothing washed with soap in hard water gets a dingy appearance due to the adherence of sticky scum in the fabric. Detergents are better cleansing agents than soap as they are not affected by hard water.

4.7 TYPES OF HARDNESS

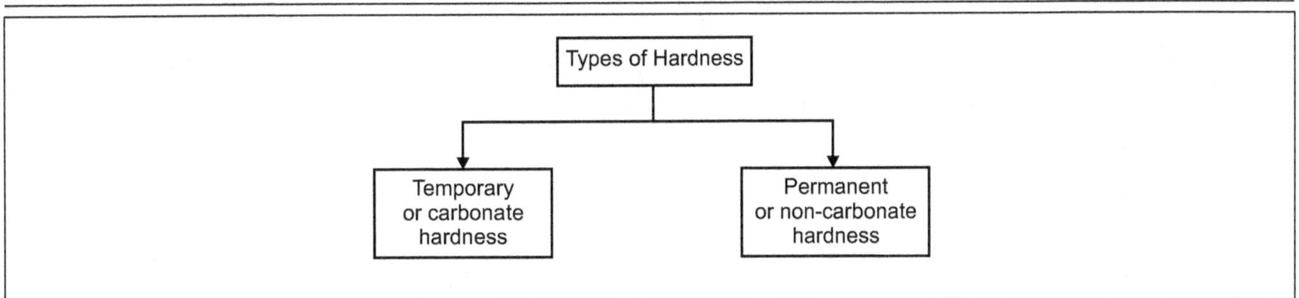

- There are two types of hardness such as :
  1. Temporary or carbonate hardness.
  2. Permanent or non-carbonate hardness.

1. Temporary or Carbonate hardness :

- It is defined as the hardness due to bicarbonates of calcium and magnesium.
- Temporary hardness is mostly removed by mere boiling of water, where bicarbonates are decomposed producing insoluble carbonates, carbon dioxide escapes as it is not soluble at higher temperature.

$$Ca(HCO_3)_2 \rightarrow CaCO_3 \downarrow + CO_2 \uparrow + H_2O$$
(Soluble)   (Insoluble)

$$Mg(HCO_3)_2 \rightarrow MgCO_3 \downarrow + CO_2 \uparrow + H_2O$$
(Soluble)   (Insoluble)

2. Permanent or Non-carbonate hardness :

- It is due to the presence of dissolved chlorides, sulphates of calcium and magnesium.
- Unlike temporary hardness, permanent hardness is not destroyed on boiling.
- It requires special chemical treatment for the removal of hardness causing salts like softening methods during which soluble salts are converted to insoluble salts.
- The softening methods used on large scale in industries are lime-soda, zeolite or permutit and ion-exchange etc.

4.7.1 Difference between Temporary and Permanent Hardness of Water

Temporary Hardness	Permanent Hardness
1. This type of hardness is due to soluble bicarbonates of calcium and magnesium.	1. It is due to soluble chloride and sulphate salts of calcium and magnesium.
2. It is due to carbonates, hence known as carbonate hardness.	2. It is due to other salts, hence known as non-carbonate hardness.
3. This type of hardness can be removed by simple process such as boiling. The name **temporary** indicates the property which is not difficult to be changed.	3. This type of hardness cannot be removed by simple techniques. Hence the term **permanent** is used to describe these characteristics.
4. Methods used to remove temporary hardness (like boiling, Clark's) cannot be used to remove permanent hardness.	4. Methods used to remove permanent hardness (permutit, ion exchange) also remove temporary hardness from water.

4.8 DEGREE OF HARDNESS

- The concentration of hardness is expressed in terms of an equivalent amount of $CaCO_3$ since this method permits the addition and subtraction of concentration of hardness. The choice of $CaCO_3$ is due to its molecular weight 100 and equivalent weight 50. Also it is insoluble in water and can be easily precipitated.

The units commonly used to express hardness of water are as follows :

(1) Parts per million (ppm) is the parts of calcium carbonate equivalent hardness per million part of water.

$$1 \text{ ppm } = 1 \text{ part of } CaCO_3 \text{ equivalent hardness per million parts of water}$$

(2) Milligrams per litre (mg/litre) is the number of milligrams of $CaCO_3$ equivalent hardness present per litre of water.

$$1 \text{ mg/litre } = 1 \text{ mg of } CaCO_3 \text{ equivalent in 1 litre}$$

Now weight of 1 litre water $= 1000 \text{ gm} = 1000 \times 100 \text{ mg} = 10^6 \text{ mg}$

$\cdot$ $1 \text{ mg/litre } = 1 \text{ mg of } CaCO_3 \text{ equivalent per } 10^6 \text{ mg of water}$

$$1 \text{ mg/litre } = 1 \text{ ppm}$$

(3) Clarke's method : Clarke's degree is the number of grains (1/7000 lb) of $CaCO_3$ equivalent per gram (10 lb) of water. i.e. one part of $CaCO_3$ equivalent per 70,000 parts of water.

$$1^\circ Cl = 1 \text{ part of } CaCO_3 \text{ equivalent per 70,000 parts of water}$$

(4) Degree french is the parts of $CaCO_3$ equivalent per 1,00,000 parts of water.

$$1^\circ Fr = 1 \text{ part of } CaCO_3 \text{ equivalent per } 10^5 \text{ parts of water}$$

$\cdot$ $1 \text{ ppm } = 1 \text{ mg/lit} = 0.1^\circ Fr = 0.07^\circ Cl$

Requisites of Industrial Water :

- The quality of water required for industries is so variable that some of them install their own water supply plant. Some industries require specific care e.g.,

  (a) Artificial silk : The water should be clear, bright and absolutely free from colour. It should be soft and should contain only a small amount of saline constituents in solution.

  (b) Drying : It should be free iron, colour and turbidity and should have smaller quantity of salts.

  (c) Paper industry : The water must be free from colour, sedimentary matter and traces of iron and manganese. The saline constituents should be present in such quantity that no appreciable amount is deposited in the dryer paper. However the water of any quality can be used for the manufacture of coarse brown paper.

  (d) Photographic films : There should be sufficient supply of pure water which is colourless and absolutely free from suspended matter and metals such as iron, zinc and manganese.

(e) Tanning : The water should be free from iron and lime salts should be minimum.

(f) Steam raising : The water should be free from sedimentary matter and should contain little free carbonic acid. The salts which can easily be removed by heating should be preferably absent. Calcium and magnesium chlorides should be absent.

4.9 BOILER AND STEAM GENERATION

- The manufacturing industries need water for a great variety of purposes, out of which steam generation is the most important one.
- Water containing dissolved salts of sulphates, bicarbonates, chlorides of calcium and magnesium, iron salts has an adverse effect on steam boilers.
- Hence, water for raising steam in boilers must be soft and must not contain too much dissolved or suspended matter; so as to avoid the troubles of (i) Corrosion, (ii) Caustic embrittlement, (iii) Priming and foaming, (iv) Scale and sludge formation.

1. **Boiler Corrosion :**
   - The most serious problem, created by the use of unsuitable water in boilers, is corrosion. *"Corrosion can be defined as the decaying of metals by a chemical or electro-chemical reaction with their environment".*
   - Water containing various types of impurities cause corrosion of the inner surface of the boiler. Corrosion in boiler is due to the following reasons viz. (i) Dissolved gases, (ii) Dissolved salts and (iii) Acidity or alkalinity of water.

2. **Caustic embrittlement (or Caustic corrosion) :**
   - This is a type of boiler corrosion, caused by the use of highly alkaline water in boiler and it is generally observed in the boiler which operates under a high pressure.
   - During water softening process, a small quantity of sodium carbonate (Na_2CO_3) is added. In high pressure boilers, sodium carbonate decomposes to give sodium hydroxide and carbon dioxide.

$$Na_2CO_3 + H_2O \rightarrow 2NaOH + CO_2\uparrow$$

   - Due to the formation of sodium hydroxide (caustic soda), water becomes more alkaline. A number of minute cracks are observed on the inner lining of boiler. This alkaline water flows into such minute cracks by capillary action. Here water evaporates and the dissolved caustic soda is left behind. The amount of caustic soda goes on increasing due to progressive evaporation. The alkaline action of caustic soda attacks the surrounding areas of cracks thereby dissolving iron material of the boiler. This causes embrittlement of boiler parts, particularly, at stressed parts like rivets, bends, joints etc. causing even a failure of the boiler.
   - Caustic embrittlement can be avoided
     (i) by using sodium phosphate instead of sodium carbonate for softening water,
     (ii) by adding tannin or lignin as additives to the boiler water, since these block the minute cracks thereby preventing infiltration of caustic soda solution,
     (iii) by adjusting the alkalinity of water to optimum level (pH – 7 to 9).

3. **Priming and Foaming :**
   - When a boiler is steaming rapidly, some particles of liquid water are mixed with the steam and pass from the boiler along with the steam. This process is known as *priming*.
   - The cause of priming is due to dissolved solids, particularly the suspended solid or high steam velocities or improper boiler design.
   - Foaming is the production of persistent foam or bubbles in boilers, which do not break easily. Foaming occurring particularly at the water surface is due to the presence of large quantity of dissolved matter in water, by which the surface tension of water is lowered. This is responsible for the phenomenon of foaming.
   - There are some disadvantages of priming process :
     (i) The actual height of water in boiler is not judged.
     (ii) To adjust the steam pressure in the boiler, more heat is required. Therefore, efficiency of the process is decreased.
     (iii) Water and dissolved salts may enter the parts of machinery which decreases the life of the machinery.

- Priming can be avoided by using mechanical steam purifier, maintaining low levels in boilers, avoiding rapid changes in steaming rate and using pure water. Foaming can be avoided by blowdown operations to predetermined intervals, adding castor oil or anti-foaming agents.

4.   Scale and Sludge Formation in Boiler :

- By continuous evaporation of water in the boiler, concentration of dissolved salts increases. When their concentration reaches to the saturation point, they are thrown out of the water in the form of precipitates on the inner walls of the boiler.

- If the precipitate is loose and slimy, it is known as *'sludge.'* On the other hand, if the precipitated matter forms a hard adhering coating on the inner walls of the boiler, it is known as *'scale.'*

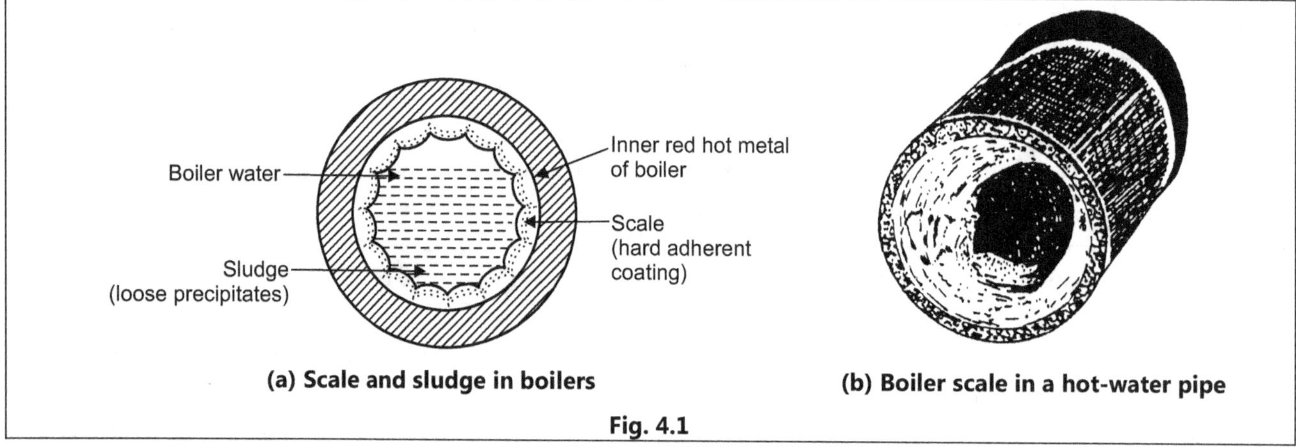

(a) Scale and sludge in boilers   **(b) Boiler scale in a hot-water pipe**

Fig. 4.1

4.10 SCALE AND SLUDGE FORMATION   (Nov. 18)

4.10.1 Scale Formation in Boilers

- *"The precipitated matter forms a hard adhering coating on the inner surface of the boiler which is called boiler scales."*
- Scales are so hard and adherent that it is difficult to remove them, even with the help of hammer or chisel. The scales are bad conductors of heat and are more troublesome.
- Scales are of following types : (i) Carbonates, (ii) Sulphates, (iii) Silicates, (iv) Phosphates, and (v) Oxides.
- The hardness of scale depends upon the nature of impurities present in the water.
- The most troublesome scales are formed due to the presence of sulphates and silicates of calcium and magnesium. Such scales are non-porous, hard, more adherent and non-conductor of heat, hence these cause overheating of the boiler material.
- Calcium carbonate forms a non-adherent scale and can be easily removed by blowing off or washing out.
- Magnesium chloride, if present in the scale will bring about corrosion of tubes and boiler plates.

4.10.2 Sludge Formation in Boilers

- *"When a soft, loose slimy deposits are formed inside the* boiler *and do not stick up permanently, they are known as sludges."*
- Sludges are generally formed at comparatively cooler parts of the boiler, and they collect in areas of the system, where the flow rate is slow or at bends in the line. In such parts they build up a deposit which choke the steam pipes.
- "Sludges are formed by substances which have greater solubilities in hot water than in cold water e.g. $CaCl_2$, $MgCl_2$, $MgSO_4$, $MgCO_3$ etc.
- Sludges are poor conductors of heat, so they tend to waste a portion of heat generated.
- The formation of sludge is not a major problem, because it can be removed by frequent blow-down operation or simply by brushing.
- Many times sludges are formed along with the scales, then these two are mixed and both get deposited as scales. Sometimes, excess of sludge formation disturbs the working of the boiler.

4.10.3 Difference between Scales and Sludges

	Scales	Sludges
1.	Form an adherent coating within the boiler which cannot be removed even by mechanical means.	Form slimy or less adherent coating within the boiler which can be removed by mechanical means.
2.	Scales are harder and more permeable.	Sludges are soft and less permeable.
3.	Scales are bad conductors of heat.	Sludges are poor conductors of heat.
4.	These are formed throughout the metal surface in contact with water.	These are formed at comparatively cooler portions of the boiler. (e.g. in bends of lines).

4.10.4 Disadvantages of Scale and Sludge Formation in Boilers

1. **Wastage of fuel :**
   - The scales are hard and bad conductors of heat. Hence, it does not allow the transfer of heat from hot plates of boiler to the water inside. Therefore in order to get a steady supply of steam more heat has to be applied. This results in a large consumption of fuel. The wastage of fuel depends upon thickness of the scale.

2. **Lowering safety of boiler :**
   - Due to scale formation, overheating of boiler is to be done in order to maintain a constant supply of steam.
   - The overheating of the boiler tube makes the boiler material softer and weaker and this causes distortion of boiler tubes (due to damage of joints and rivets) and make the boiler unsafe to bear the pressure of the steam, especially in high pressure boilers.

3. **Danger of explosion :**
   - When thick scales crack due to uneven expansion of scale and boiler material, the water comes suddenly in contact with overheated iron plates of boiler.
   - This causes the formation of a large amount of steam suddenly inside the boiler. Hence, sudden high pressure is developed, which may cause explosion of the boiler.

4. **Decrease in efficiency :**
   - Scales may sometimes deposit in the valves and condensers of boiler and choking them partially. This results in decrease in efficiency of the boiler.

5. **Shortening the life of boiler :**
   - The life of a boiler is shortened due to the following reasons :
   - (a) The steam reacts with red hot iron plates of the boiler forming non-adherent iron oxide and liberates the hydrogen gas. This causes thinning of the boiler plates with the continuous reaction of steam.

$$3Fe + 4H_2O \longrightarrow Fe_3O_4 + 4H_2 \uparrow$$

   - (b) Magnesium chloride, if present in the scale, will bring about corrosion of tubes and boiler plates like a chain reaction producing HCl again and again.

$$MgCl_2 + 2H_2O \rightarrow Mg(OH)_2 \downarrow + 2HCl$$
$$Fe + 2HCl \rightarrow FeCl_2 + H_2 \uparrow$$
$$FeCl_2 + 2H_2O \rightarrow Fe(OH)_2 + 2HCl$$

4.10.5 Removal of Scales and Sludges

   - Since the scales and sludges are harmful, they should be removed from time to time.
   - (i) With the help of a scraper, knife, blade or a piece of wood or by wire brushing, if the scales are loosely adhering.
   - (ii) By giving thermal shocks (i.e. heating the boiler and then cooling suddenly with cold water), if they are brittle.
   - (iii) By dissolving them in some chemicals, if they are hard and adherent. Thus, calcium carbonate scale can be removed by using 5 - 15 % HCl, calcium sulphate scale can be dissolved by adding EDTA (Ethylene Diamine Tetra Acetic acid).
   - (iv) By frequent blow-down operation, if the scales are loosely adhering.

4.11 WATER SOFTENING

- The process of removing soluble calcium and magnesium salts from hard water is called softening the water.

- During the process of softening of hard water, the soluble salts are converted into insoluble salts.

- These insoluble salts can be removed by filtration and soft water can be obtained.

- Following methods are used for softening of water :

    (1) Lime – soda process : (a) Hot lime-soda process. (b) Cold lime-soda process.

    (2) Zeolite or permutit process.

    (3) Ion exchange process.

4.11.1 Clark's Method of Water Softening

- **Lime softening**, also known as Clark's process, is a type of water treatment used for water softening which uses the addition of lime water (calcium hydroxide) to remove hardness (calcium and magnesium) ions by precipitation. The process is also effective at removing a variety of microorganisms and dissolved organic matter by flocculation.

Chemistry :

- A lime in the form of limewater is added to raw water, the pH is raised and the equilibrium of carbonate species in the water is shifted. Dissolved carbon dioxide (CO_2) is changed into bicarbonate (HCO_3^-) and then carbonate (CO_3^{2-}). This action causes calcium carbonate to precipitate due to exceeding the solubility product. Additionally, magnesium can be precipitated as magnesium hydroxide in a double displacement reaction.

- In the process both the calcium (and to an extent magnesium) in the raw water as well as the calcium added with the lime are precipitated. This is in contrast to ion exchange softening where sodium is exchanged for calcium and magnesium ions. In lime softening, there is a substantial reduction in total dissolved solids (TDS) whereas in ion exchange softening (sometimes referred to as zeolite softening), there is no significant change in the level of TDS.

- Lime softening can also be used to remove iron, manganese, radium and arsenic from water.

Future uses :

- Lime softening is now often combined with newer membrane processes to reduce waste streams. Lime softening can be applied to the concentrate (or reject stream) of membrane processes, thereby providing a stream of substantially reduced hardness (and thus TDS), that may be used in the finished stream. Also, in cases with very hard source water (often the case in Midwestern USA ethanol production plants), lime softening can be used to pre-treat the membrane feed water.

4.11.2 Lime-Soda Process

- In this process, lime $Ca(OH)_2$ (calcium hydroxide) and soda Na_2CO_3 (sodium carbonate) are used to soften the water.

- By this method, both temporary and permanent hardness of water can be removed. Bicarbonates and carbonates are removed by lime while sulphates and chlorides are removed by soda.

$$Ca(HCO_3)_2 + Ca(OH)_2 \rightarrow 2\,CaCO_3 \downarrow + 2H_2O$$

$$Mg(HCO_3)_2 + Ca(OH)_2 \rightarrow CaCO_3 \downarrow + MgCO_3 \downarrow + 2H_2O$$

$$MgCO_3 + Ca(OH)_2 \rightarrow CaCO_3 \downarrow + Mg(OH)_2$$

- If water contains sulphates and chlorides of magnesium then both soda and lime are required.

$$MgSO_4 + Na_2CO_3 + Ca(OH)_2 \rightarrow Mg(OH)_2 + CaCO_3 \downarrow + Na_2SO_4$$

$$MgCl_2 + Na_2CO_3 + Ca(OH)_2 \rightarrow Mg(OH)_2 + CaCO_3 \downarrow + 2NaCl$$

- On the other hand, if the sulphates and chlorides of calcium are present then only soda-ash is required.

$$CaSO_4 + Na_2CO_3 \rightarrow CaCO_3 \downarrow + Na_2SO_4$$

$$CaCl_2 + Na_2CO_3 \rightarrow CaCO_3 \downarrow + 2NaCl$$

- Lime soda process can be carried out either in cold water or in hot water. If the process is carried out in cold water, it is called as *cold-lime soda* process. If it is carried out in hot water then, it is known as *hot-lime soda* process.

1. **Cold-Lime Soda Process :**

    - In this method, calculated quantity of lime and soda are mixed with water at room temperature.

    - The precipitate obtained is finely divided and to coagulate the precipitate, a small quantity of coagulant such as alum should be used. Removal of precipitate by filtration gives soft water.

    - The process can be carried out by two methods viz., batch process and continuous process.

(a) **Batch Process :**

    - A calculated quantity of lime and soda-ash is added to water in a large tank made up of steel plates. It is provided with a mechanical stirrer for stirring or agitating the mixture. (Refer Fig. 4.2).

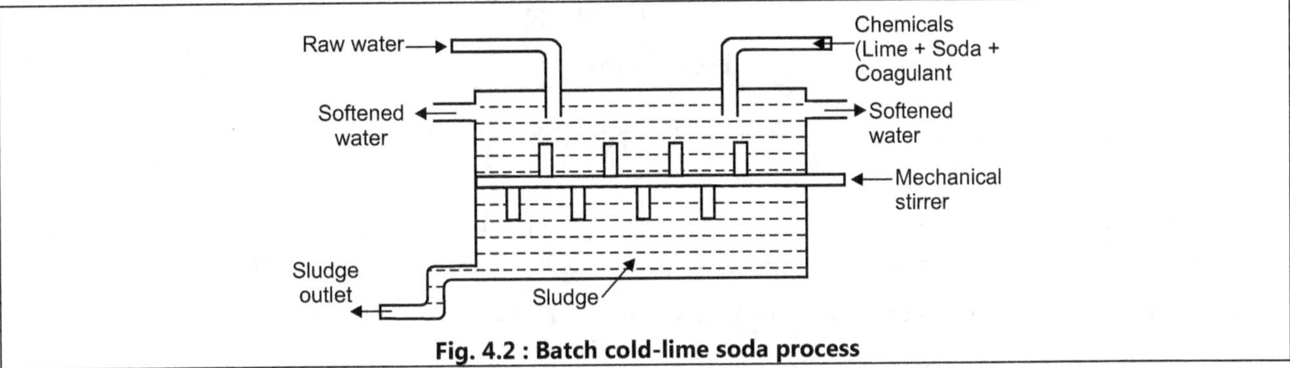

Fig. 4.2 : Batch cold-lime soda process

    - After thorough mixing, mechanical stirring is stopped and a small quantity of coagulant (alum) is added to coagulate the precipitate of impurities.

$$Al_2(SO_4)_3 + 3Ca(HCO_3)_2 \rightarrow 2Al(OH)_3 \downarrow + 3CaSO_4 + 6CO_2 \uparrow$$

    - The precipitated impurities i.e. sludge settle down at the bottom of the tank.

    - The water is pumped out and allowed to pass through a filter unit of sand and coal. After filtration the soft water is obtained.

    - To complete the process, 2 to 3 hours are required. After removal of sludges the tank is washed with water.

    - The same procedure described above is repeated again to soften more quantity of hard water.

(b) **Continuous Process :**

    - This process is a continuous process and therefore, time is not wasted during operations.

    - In this process, hard water and calculated quantity of chemicals (lime + soda + coagulant) are fed from the top into the inner vertical circular chamber fitted with a vertical rotating shaft carrying a number of paddles.

    - As the hard water and chemicals flow down, there is a vigorous stirring and continuous mixing takes place.

    - The softened water, as it comes into the outer coaxial chamber, rises upwards. The heavy sludge (or precipitated flock) settles down in the outer chamber by the time the softened water rushes up.

    - The softened water then passes through a filtering medium (usually made of wood fibres to ensure complete removal of sludge).

    - Filtered soft water finally flows out continuously through the outlet nearly at the top.

    - The sludges at the bottom are removed through sludge outlet occasionally.

    - This process provides water containing hardness of 50 to 60 ppm.

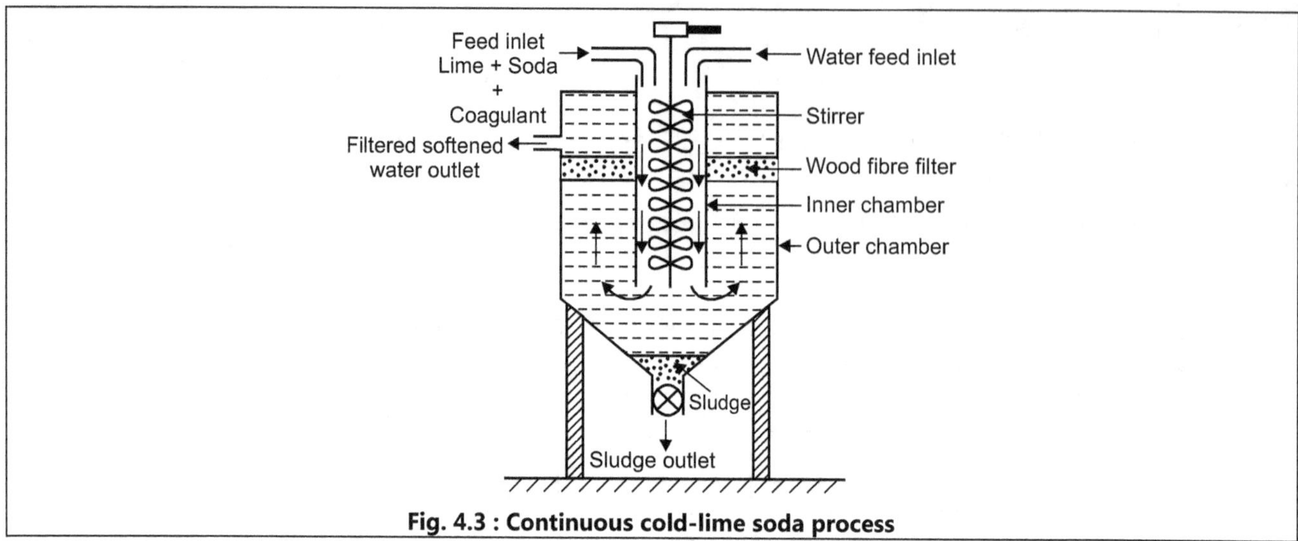

Fig. 4.3 : Continuous cold-lime soda process

2. Hot-Lime Soda Process :

- In this process, hard water is treated with lime-soda at a temperature of 80 to 150°C. Since the process is carried out at an elevated temperature, the speed of the reaction increases and the process is completed in 15 minutes.

- There are certain advantages of hot-lime soda process over cold-lime soda process.

(i) The rate of precipitating reaction is faster at elevated temperature, hence, the reaction is completed in a short time of 15 minutes.

(ii) The coagulant is not needed for precipitation, because the precipitate settles rapidly.

(iii) The dissolved gases such as CO_2 are removed at higher temperature.

(iv) The water obtained by this process is much softer as compared with cold-lime soda process.

(v) The process of filtration becomes much faster and easier.

(vi) The softening capacity of this process is much higher than the cold-lime soda process.

(a) Batch Process :

- Hard water is taken in a rectangular tank made up of steel plates, which is provided by heating coils.

- Calculated quantity of lime and soda is added to the tank. By increasing the temperature of water to about 150°C, the speed of the reaction can be increased. Within 15 minutes the process is completed.

- By using a water pump the clear and soft water is pumped out and allowed to pass through a filter unit. Sludges are removed from the tank. The tank is washed thoroughly and the same procedure is repeated again.

(b) Continuous Process :

- The apparatus consists of a tank with two concentric cylindrical chambers.

- The lower end of the inner chamber is funnel-shaped and consists of three inlets.

- The outer chamber is big with larger cross section at the top than at the bottom.

- This facilitates rise of water from the bottom. The hard water and chemicals are added through the left side inlets and from the third inlet, steam is allowed to pass to increase the temperature of the water.

- The precipitate formed settles at the bottom and soft clear water is withdrawn from the top of the outer chamber. The hardness of the water is reduced to 25 ppm from this method.

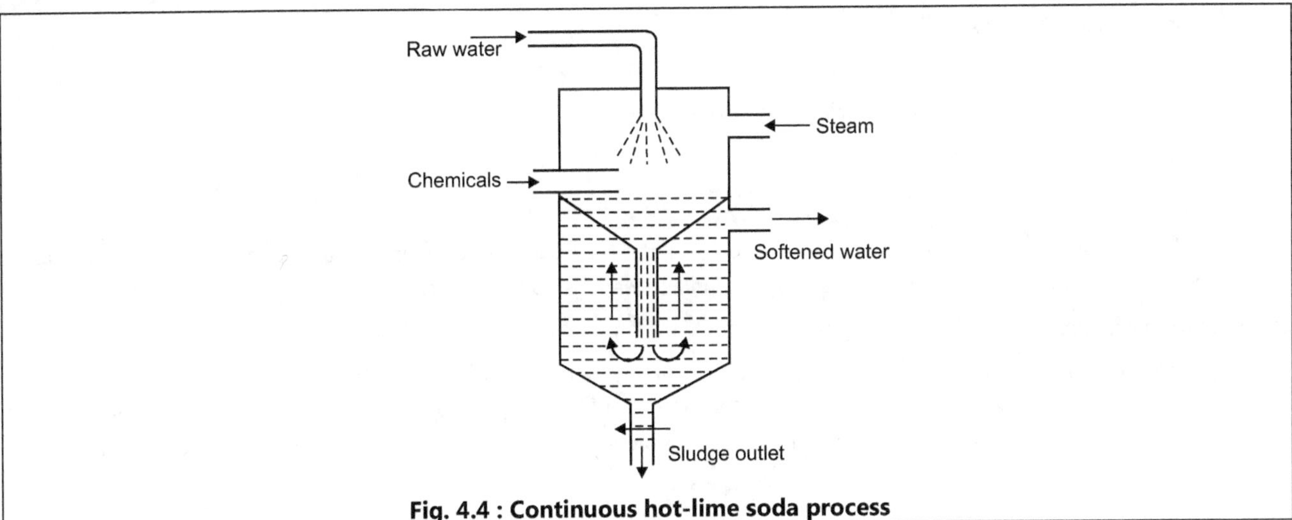

Fig. 4.4 : Continuous hot-lime soda process

4.11.3 Zeolite or Permutit Process

- This is the process used for removing both temporary and permanent hardness of water.

- *Permutit or zeolites are complex silicates of several metallic and non-metallic oxides. They have an approximate chemical formula ($Na_2Al_2Si_2O_8 \cdot 6H_2O$). These silicates hold sodium ions loosely, hence these are called as sodium permutit (Na_2P)* or sodium zeolite (Na_2Z) where permutit and zeolite stand for $Al_2Si_2O_8 \cdot 6H_2O$.

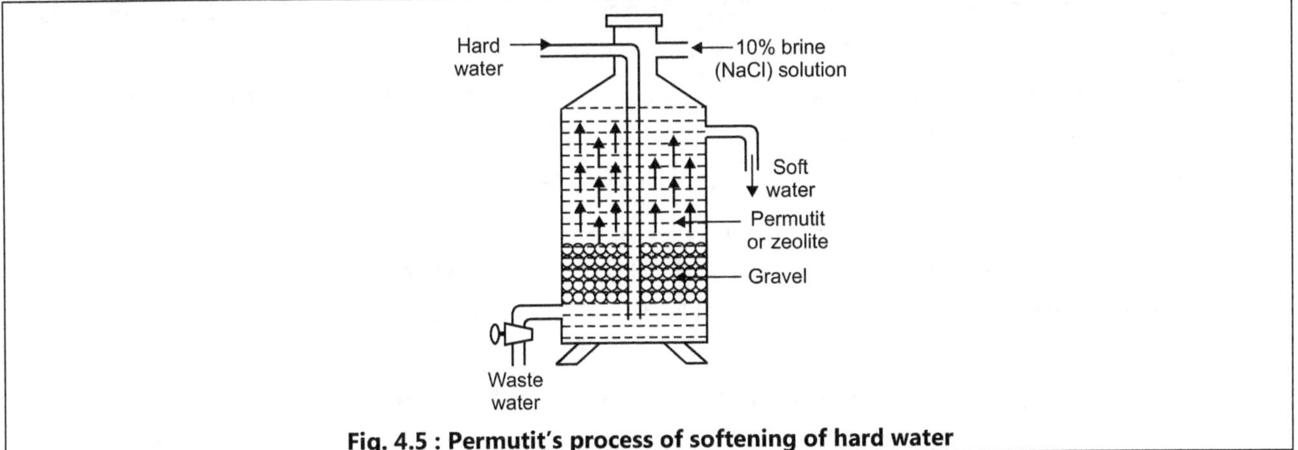

Fig. 4.5 : Permutit's process of softening of hard water

Principle :

- Sodium permutit (Na_2P) when comes in contact with hard water, it exchanges its Na^+ ions with Ca^{++} and Mg^{++} ions to form insoluble calcium permutit (CaP) and magnesium permutit (MgP).

Procedure :

- Sodium permutit is placed in a suitable container and hard water is allowed to pass through it (Refer Fig. 4.5)

- The calcium and magnesium salts react with it forming insoluble calcium and magnesium permutits. These salts are retained in the filter bed and water after reaction is therefore, free from calcium and magnesium salts. Only harmless sodium salts are left in the water.

- The following reactions take place in the process :

  For temporary hardness :

$$Ca(HCO_3)_2 + Na_2P \rightarrow CaP \downarrow + 2NaHCO_3$$

$$Mg(HCO_3)_2 + Na_2P \rightarrow MgP \downarrow + 2NaHCO_3$$

For permanent hardness :

$$CaCl_2 + Na_2P \rightarrow CaP \downarrow + 2NaCl$$
$$MgCl_2 + Na_2P \rightarrow MgP \downarrow + 2NaCl$$
$$CaSO_4 + Na_2P \rightarrow CaP \downarrow + Na_2SO_4$$
$$MgSO_4 + Na_2P \rightarrow MgP \downarrow + Na_2SO_4$$

- When the process is continued for about 12 hours, all the Na^+ ions from the permutit are replaced by Ca^{++} and Mg^{++} ions and it is found that the permutit stops working and water is no more softened.

Regeneration reaction :

- When the permutit is exhausted (i.e. it is completely converted into CaP and MgP), it is regenerated by treating with 10 % brine (NaCl) solution for a few minutes, sodium permutit (Na_2P) is formed and can again be used for softening of more hard water. The reactions in regeneration can be shown as follows :

$$CaP + 2NaCl \rightarrow Na_2P + CaCl_2$$
$$MgP + 2NaCl \rightarrow Na_2P + MgCl_2$$

Limitations of Permutit or Zeolite Process :

(i) If hard water contains large quantities of coloured ions like manganese ions (Mn^{++}) or ferrous ions (Fe^{++}), then these ions form manganese or iron permutit. Such permutit cannot be regenerated easily.

(ii) If hard water contains mineral acids, then it may destroy permutits. Moreover, in zeolite process only cations like Ca^{++}, Mg^{++} are replaced but anions like CO_3^{--}, HCO_3^{-} etc. remain in water. Such water cannot be used in boilers as CO_2 is released from it and is extremely corrosive to boiler material.

(iii) If hard water is turbid, then turbidity may clog (or block) the pores of permutit and it restricts the flow of water.

Advantages of Permutit or Zeolite Process :

(i) The water having zero hardness can be obtained i.e. it removes the hardness completely.

(ii) The equipment used is compact, thus occupying a small space.

(iii) It is a clean process as no impurities are precipitated and hence, there is no sludge formation.

(iv) The process automatically adjusts itself for different hardness of incoming water.

(v) It requires less time.

4.11.4 Ion-Exchange (or Deionisation or Demineralisation Process)

- This is the modern development in water softening method.

- Certain organic compounds possess a property like zeolite i.e. they are capable of exchanging ions. Such organic synthetic compounds are known as resins.

- There are two types of resins : (1) Cation exchange resins, (2) Anion exchange resins.

(a) Cation exchange resins : These resins are capable of exchanging rapidly cations by H^+ ions. Cation exchange resins can be represented as RH_2, so their exchange reaction with cations (e.g. Ca^{++}) is :

$$RH_2 + Ca^{++} \rightarrow RCa + 2H^+$$

(b) Anion exchange resins : These resins are capable of exchanging rapidly anions by OH^- ions. Anion exchange resins can be represented as $R'(OH)_2$, so their exchange reaction with anions (e.g. SO_4^{--}) is :

$$R'(OH)_2 + SO_4^{--} \rightarrow R'SO_4 + 2OH^-$$

- From the above it is clear that, if hard water is passed first through cation exchanger and then through anion exchanger, the resulting water will be free from both cations and anions and water is said to be *'deionised'* or *'demineralised.'*

Process :

- The process of ion exchange is carried out as follows :

(i) It consists of two cylindrical towers, out of which the first tower consists of 'cation exchanger' (RH_2) and the other one consists of 'anion exchanger' $[R'(OH)_2]$ (Refer Fig. 4.6).

(ii) Hard (or impure) water is first allowed to pass through a tower containing cation exchanger, which removes all the cations like Ca^{++}, Mg^{++}, Na^+, Fe^{++} etc. and releases H^+ ions.

$$RH_2 \quad + \quad Ca^{++} \quad \rightarrow \quad RCa \quad + \quad 2H^+$$
$$RH_2 \quad + \quad Mg^{++} \quad \rightarrow \quad RMg \quad + \quad 2H^+$$
$$RH_2 \quad + \quad 2Na^+ \quad \rightarrow \quad RNa_2 \quad + \quad 2H^+$$

(iii) Thus, the anions like chlorides, sulphates and bicarbonates are converted into their corresponding acids HCl, H_2SO_4 and H_2CO_3. In other words, water from cation exchanger is free from all cations, but it is acidic.

(iv) This acidic water is then passed through another tower containing an anion exchanger, where acids are converted into water.

$$R'(OH)_2 \quad + \ 2HCl \quad \rightarrow \quad R'Cl_2 \quad + \quad 2H_2O$$
$$R'(OH)_2 \quad + \ H_2SO_4 \quad \rightarrow \quad R'SO_4 \quad + \quad 2H_2O$$
$$R'(OH)_2 \quad + \ H_2CO_3 \quad \rightarrow \quad R'CO_3 \quad + \quad 2H_2O$$

(v) Consequently, water thus produced is free from all ions (cations and anions) and is virtually distilled water.

(vi) The water is finally freed from dissolved gases like CO_2, etc. by passing it through a 'degasifier', which is a tower whose sides are heated and which is connected to a vacuum pump. High temperature and low pressure reduce the quantity of dissolved CO_2 and O_2 in water. Such softened water can be used for industrial purposes.

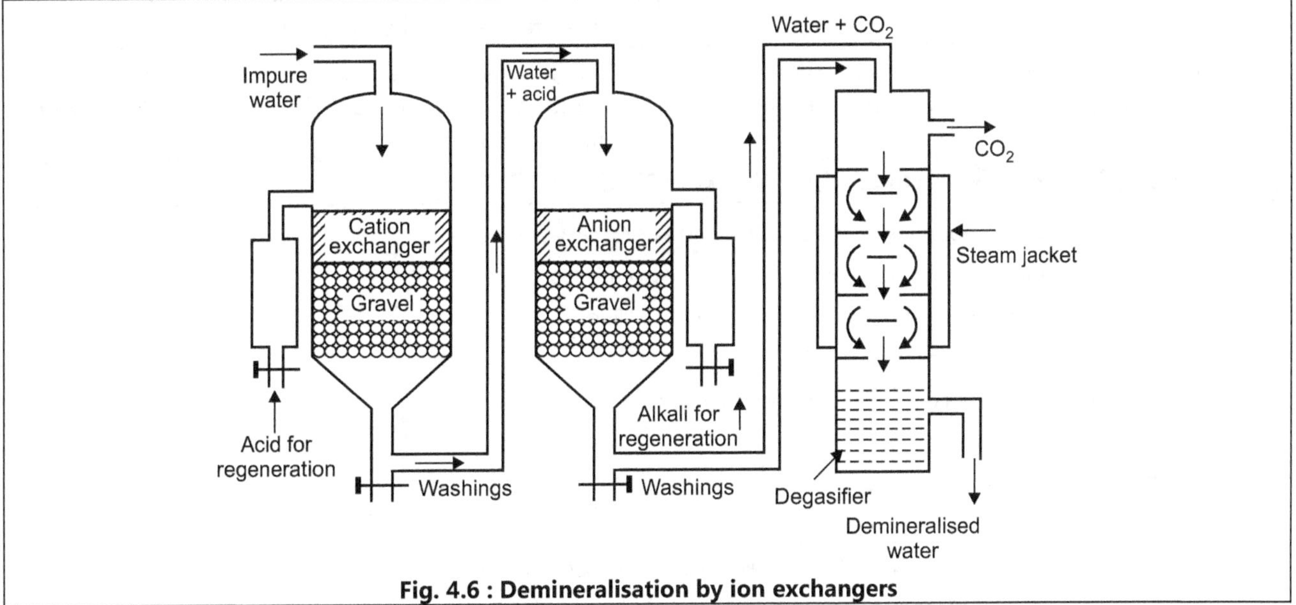

Fig. 4.6 : Demineralisation by ion exchangers

Regeneration :

- When the resins get exhausted (i.e. when their capacity to exchange H^+ and OH^- ions is lost) they are regenerated.

- The exhausted cation exchanger is then regenerated by passing a dilute solution of acid through the first tower (e.g. dil. HCl or dil. H_2SO_4).

$$RCa \quad + \ 2HCl \quad \rightarrow \quad RH_2 \quad + \quad CaCl_2$$
$$RMg \quad + \ 2HCl \quad \rightarrow \quad RH_2 \quad + \quad MgCl_2$$

- The washing containing $CaCl_2$, $MgCl_2$ (or $CaSO_4$, $MgSO_4$) etc. is passed to drain.

- Similarly, the exhausted anion exchanger is regenerated by passing dilute solution of alkali through the second tower (e.g. dil NaOH or KOH).

$$R'Cl_2 \ + \ 2NaOH \ \rightarrow \ R'(OH)_2 \ + \ 2NaCl$$

$$R'SO_4 \ + \ 2NaOH \ \rightarrow \ R'(OH)_2 \ + \ Na_2SO_4$$

- Thus washings containing NaCl, Na_2SO_4 etc. are also passed to drain. The regenerated resins are then used again.

4.11.5 Difference between Ion-exchange Process and Zeolite Process

Ion-exchange process	Zeolite process
1. This process can produce softened water with residual hardness ranging between 0 to 2 ppm.	1. This process can produce softened water with residual hardness ranging between 0 to 15 ppm.
2. The resultant water is suitable for all types of boilers, especially high pressure boilers.	2. The resultant water is not suitable for use in high pressure boilers. Water can be used only in low or medium pressure boilers.
3. The cation and anion exchange beds used are more expensive. Hence, capital cost is high.	3. Zeolite softener is comparatively cheap, hence capital cost is lower.
4. The softening plant is not compact, hence occupies more space.	4. The softening plant is compact. Hence occupies less space.
5. The process effectively removes all the hardness causing substances. It can also remove alkali metals such as Na or K, as chlorides or sulphates completely.	5. This process can remove only Ca^{+2}, Mg^{+2}, Fe^{+2} and Mn^{+2} ions. Hence, softened water contains salts like NaCl, $NaHCO_3$, Na_2SO_4 etc. in dissolved form.
6. This process is useful for acidic as well as alkaline water.	6. This process is not useful for highly acidic water as acids affect zeolite bed, because zeolites get dissolved in.
7. Soft de-ionized water does not cause caustic embrittlement in boilers as it is free from Na^+ ions.	7. Soft water is not suitable for boilers due to the presence of $NaHCO_3$, which subsequently forms NaOH, causing thereby *caustic embrittlement* in boilers.

4.12 PLUMBO SOLVENCY (POTABLE WATER TREATMENT)

- Water which is suitable for drinking purpose is called as potable water.

- Drinking or potable water should satisfy the following essential conditions :
    1. Water should be clear, colourless, odourless and sparkling.
    2. It should be pleasant in taste.
    3. It should be free from disease producing micro-organisms.
    4. It should be reasonably soft.
    5. Its turbidity should not be more than 10 ppm.
    6. Its dissolved solids should not be more than 500 ppm.

- Generally, for domestic supply, surface water used which is contaminated with large number of impurities such as organic matter, suspended particles, colloidal particles (turbidity), germs and bacterias.

- Therefore, to make it safe for drinking purposes, following methods can be carried out :

    (1) Sedimentation, (2) Coagulation, (3) Filtration, (4) Sterilization.

4.12.1 Sedimentation

- *"Sedimentation is the process of removing suspended impurities by allowing the water to stay undisturbed for some time in large tanks when most of the suspended particles settle down due to the force of gravity."* The clear water can be taken out from the tank with the help of pumps. The detention period in a settling tank may range from an hour to several days.

- The process of sedimentation is generally carried out in 'continuous flow type tanks' in which water flows continuously in a horizontal, or vertical direction at a slow and uniform speed. Due to gravitational force, the suspended particles get settled down at the bottom of the tank, from where they can be removed periodically. Sedimented water is taken out continuously.

4.12.2 Coagulation

- *"Coagulation is the process of removing colloidal (or fine sized) particles from water by the addition of certain chemicals known as coagulants before sedimentation."* This process is usually carried out along with sedimentation.

- Actually the colloidal (or fine sized) particles present in water either do not settle down at all or take a very long time.

- In order to have a quick settling of these particles, certain chemicals known as coagulants are used. The commonly used coagulants are the salts of iron and aluminium e.g. alum [K_2SO_4, $Al_2(SO_4)_3$, $24H_2O$], ferrous sulphate ($FeSO_4$, $7H_2O$) etc.

- These coagulants react with bicarbonates present in water, and form bulky gelatinous precipitate called 'flock'. As these flocks descend through water, they absorb or catch suspended fine particles from water and forming bigger flocks, which settle down quickly.

- The addition of coagulants to water also removes colour, odour and improves its taste.

$$Al_2(SO_4)_3 + 3Ca(HCO_3)_2 \rightarrow 2Al(OH)_3\downarrow + 3CaSO_4 + 6CO_2\uparrow$$

(Coagulant)　　Calcium　　　　　(Bulky gelatinous flock)
　　　　　　　bicarbonate

$$FeSO_4 + Mg(HCO_3)_2 \rightarrow Fe(OH)_2\downarrow + MgCO_3 + H_2O + CO_2\uparrow$$

(Coagulant)　　Magnesium　　　(Ferrous hydroxide)
　　　　　　　bicarbonate

$$4Fe(OH)_2 + 2H_2O + O_2 \rightarrow 4Fe(OH)_3\downarrow$$

　　　　　　　(Dissolved　　　　Ferric hydroxide
　　　　　　　oxygen)　　　　　(Heavy flock)

As per other concept :

- Colloidal particles are very small sized (10^{-4} to 10^{-7} cm) particles possessing either positive or negative charge. Due to similar charge, they repel one another and do not come together. Therefore, they do not settle down during sedimentation.

- Colloidal particles of clay possess negative charge. When alum (coagulant) is added to water, it provides positive aluminium ions Al^{3+}. These positively charged aluminium ions neutralise negative charge of colloidal particles of clay. As the charge is removed from the clay particles, they come nearer and combine to form particles, which settle down at the bottom of the container due to gravitational force.

4.12.3 Filtration

- *"Filtration is a process of removing insoluble colloidal and bacterial impurities by passing water through a bed of proper sized material."*

- By filtration, suspended matter, insoluble colloidal matter, most of the bacterias, colours and odour of water are removed.

- Two types of filters are commonly used in domestic water treatment :

	(i) Gravity sand filter, (ii) Pressure filter.

(i) Gravity sand filter :

Construction :

- These filters are best suited for municipal water supply.

- It consists of a large shallow rectangular tank made of concrete (Refer Fig. 4.7).

- At the bottom of the tank, there is a channel of bricks through which filtered water goes out. Over this channel, a layer of coarse and fine gravels (about 30 cm thick) and then a layer of coarse sand (about 20 cm thick) and finally a layer of fine sand (about 50 cm thick) are placed.

Working :

- Sedimented water enters the sand filter from the top. As the water percolates through the sand bed, fine suspended particles, most of the germs and bacterias are retained by the top sand layer. Clean filtered water is collected in the under drain channel, from where it is drawn out.

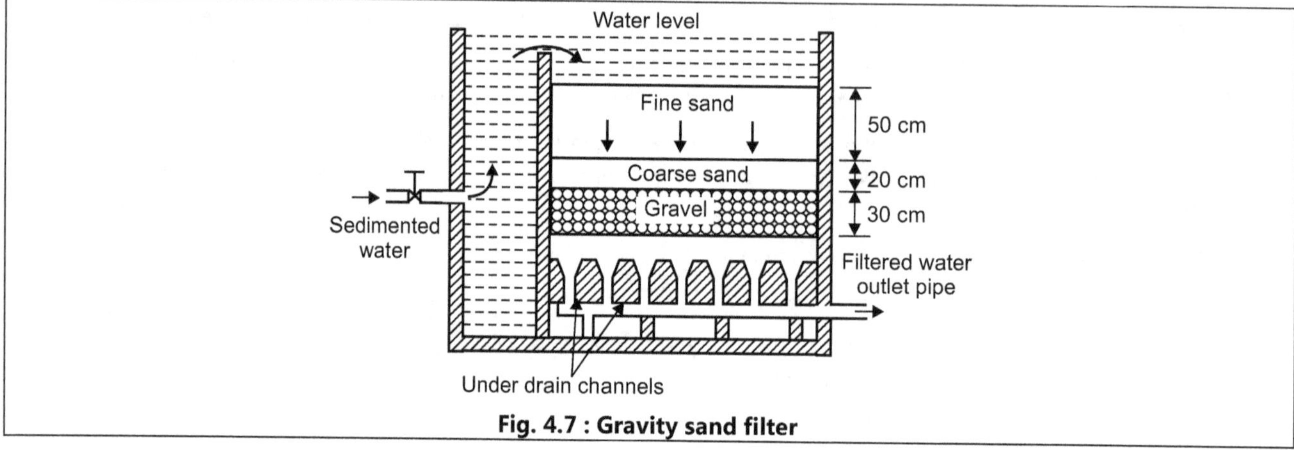

Fig. 4.7 : Gravity sand filter

Cleaning :

- The rate of filtration, after 24 hours of use becomes slow due to clogging of pores of the top sand layer by the impurities retained in the pores.

- Therefore, the portion of the top fine sand layer is scrapped off and replaced by a new sand layer. The filter is put to use again.

(ii) Pressure filter :

Construction :

- Pressure filters are economical only for small scale use and not for water supplies such as municipal water supply.

- It consists of a cylindrical vertical steel tank containing three layers of filtering media one above the other.

    (a) Pebbles (10 – 35 mm grain size) layer,

    (b) Coarse sand (5 – 7 mm grain size) layer,

    (c) Fine sand (1 – 2 mm grain size) layer.

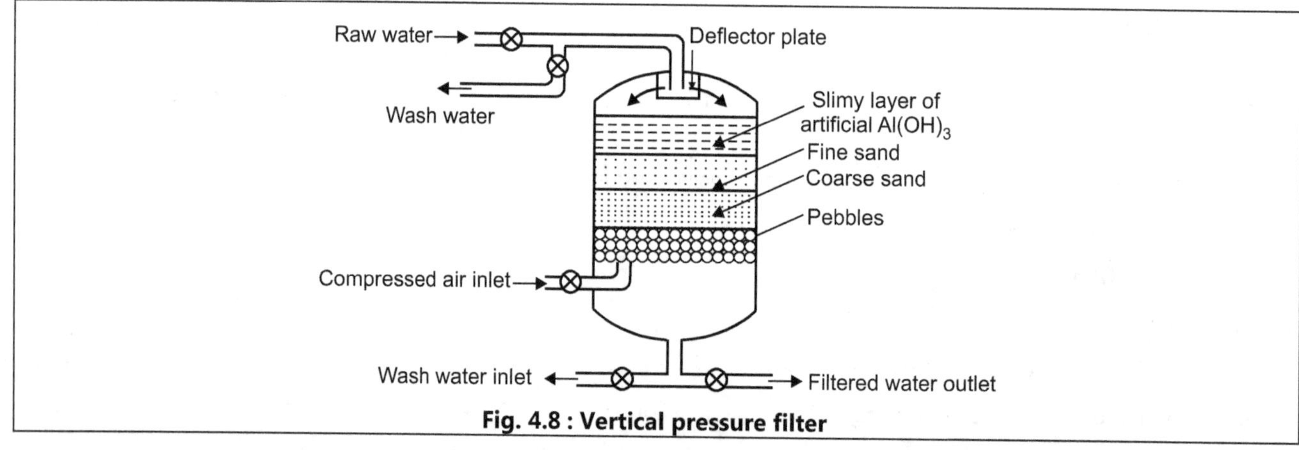

Fig. 4.8 : Vertical pressure filter

Working :

- Impure sedimented water is mixed with a small amount of alum solution. It is then forced in under pressure from the top of the tank (Refer Fig. 4.8).
- Added alum forms an artificial slimy layer on the filter bed and this helps in removal of colloidal and bacteriological impurities.
- The function of the deflector plate provided at the top is to distribute the slimy layer uniformly over the top of the filter bed.
- Filtered water as it comes out from the bottom of filter is under pressure and can thus be pumped directly.

Cleaning :

- Pressure filter gets clogged after its use for some time and its periodical cleaning is necessary. This is done by 'back washing.'
- The water under pressure is forced up from the tank which removes the matter has clogged the sand pores. The back wash water is then allowed to go to drain.
- Now-a-days compressed air is used to agitate sand before back washing.

4.12.4 Sterilization (Removal of Bacteria and Micro-organisms)

- Water after passing through sedimentation, coagulation and filtration operations still contain a small percentage of pathogenic (disease producing) bacteria. Therefore, it is necessary to remove these bacteria and micro-organisms from water.
- "The process of destroying these disease causing bacteria and micro-organisms etc. from water is known as disinfection or sterilization of water."
- The chemicals used for sterilization are known as 'sterilizers.' The disinfection of water can be carried out by the following methods :

(i) Boiling	(ii) Chlorination
(iii) Ozonisation	(iv) Aeration
(v) Ultraviolet rays	(vi) Removal of algae.

(i) Boiling :

- Sterilization strictly speaking is the boiling of water before using for domestic purposes.
- Boiling kills all the disease causing germs and bacteria during epidemics like cholera and typhoid within five minutes.
- But this method is useful only for household purposes, because the process is very expensive for municipal town supply of water.

(ii) Chlorination :

- The chlorination can be carried out by the following methods :
  - (a) By using chlorine gas
  - (b) By adding bleaching powder.
  - (c) By using chloramine.

(a) By using chlorine gas (Cl_2) :

- Chlorine can be used directly as a gas or as chlorine water for sterilization of municipal water supply.
- It reacts with water to form hypochlorous acid and nascent oxygen, both are powerful germicides.

$$Cl_2 + H_2O \rightarrow HOCl + HCl$$
$$\text{(Hypochlorous acid)}$$

$$HOCl \rightarrow HCl + [O]$$
$$\text{(Nascent oxygen)}$$

$$\text{Germs} + [O] \rightarrow \text{Germs are oxidised}$$

- However, excess of chlorine should be avoided because it produces unpleasant odour, taste and irritating effect on mucous membrane. The treated water should not contain more than 0.1 - 0.2 ppm of free chlorine.

(b) By adding bleaching powder ($CaOCl_2$) :

- Bleaching powder is a good sterilizer for small water works.

- In practice, about 1 kg of bleaching powder per 1000 litres of water is mixed and the resulting solution is allowed to stand for several hours. Hypochlorous acid (HOCl) and nascent oxygen (O) produced by the action of water on bleaching powder are powerful germicides.

$$CaOCl_2 + H_2O \rightarrow Ca(OH)_2 + Cl_2$$
$$Cl_2 + H_2O \rightarrow HOCl + HCl$$
$$HOCl \rightarrow HCl + [O]$$
$$\text{(Nascent oxygen)}$$

- The nascent oxygen liberated, oxidises germs and other harmful bacteria.

- Only calculated quantity of bleaching powder should be used, because excess of it will give a bad taste and disagreeable smell to the water.

Disadvantages :

- The use of bleaching powder has the following disadvantages :

1. Bleaching powder introduces calcium in water thereby making it more hard.

2. Bleaching powder deteriorates due to its continuous decomposition during storage. Hence, before its use, it should be analysed for its effective chlorine content.

3. If used in an excess amount, it imparts a bad taste and disagreeable smell to water.

(c) By using chloramine ($ClNH_2$) :

- Chlorine and ammonia are mixed in the ratio 2 : 1 by volume to produce a compound known as chloramine. This process is known as *'chloramination.'*

$$Cl_2 + NH_3 \rightarrow ClNH_2 + HCl$$
$$\text{(chloramine)}$$

- Chloramine is a quite stable compound and does not impart any disagreable smell and bad taste to water, hence it is considered as a better germicide than chlorine alone.

$$ClNH_2 + H_2O \rightarrow HOCl + NH_3 \uparrow$$
$$HOCl \rightarrow HCl + [O] \text{ (Nascent oxygen)}$$

Advantages of chloramination :

1. It removes irritating smell due to excess of chlorine.

2. It imparts good taste to water.

3. It checks the dissipation of chlorine (by stabilizing chlorine) when water is exposed to atmosphere, especially to sunlight.

- The chloramine tablets are used in the army for stabilizing the water. Water is collected in individual bottles in which chloramine tablets are added followed by sodium thiosulphate which removes excess of chlorine.

(iii) Ozonisation :

- This is an effective method of sterilization of water. Ozone (O_3) is unstable and it decomposes into molecular oxygen (O_2) and nascent oxygen (O).

$$O_3 \rightarrow O_2 + [O]$$
$$\text{(Nascent oxygen)}$$

- The nascent oxygen thus produced is very effective for killing all the germs and bacteria.

- In ozonisation, water is allowed to percolate through a tower having perforated partition (Refer Fig. 4.9). Ozone is allowed to enter from the bottom which kills the germs when they come in contact with water. Sterilized water is collected at the bottom of the tank.

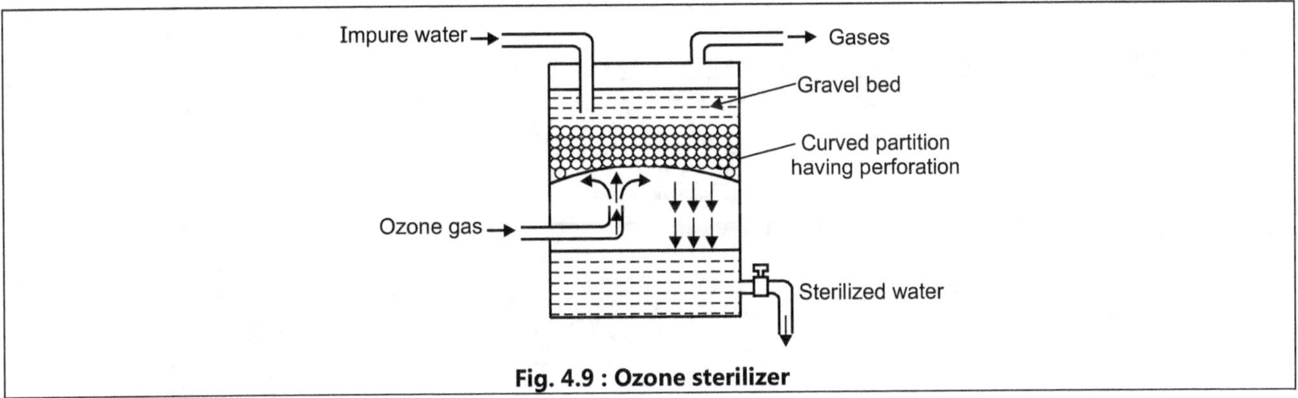

Fig. 4.9 : Ozone sterilizer

- This method is quite expensive and hence it is not employed for sterilization of municipal water supply.

Advantages of Ozonisation :
1. Ozone acts not only as a sterilizing agent but also as a bleaching, decolourising and deodourising agent.
2. It improves the taste of water.
3. In excess it is not harmful, since it is unstable and decomposes into oxygen.

(iv) Aeration :

- *"The process of spraying water in the form of fine droplets into the atmosphere is known as aeration."*
- This is the most modern method of purifying water for town supply. In this method, water is forced under pressure through a perforated pipe (Refer Fig. 4.10). As water sprays into air, it comes in intimate contact with the oxygen of air and is exposed to the ultra-violet rays of the sun. This kills the bacteria and the oxygen oxidises organic matter present in the water. It removes colour and odour also. The pure water is collected in a shallow tank placed below.
- It should be noted that as the exposure time is too short, the desirable effect may not be obtained.

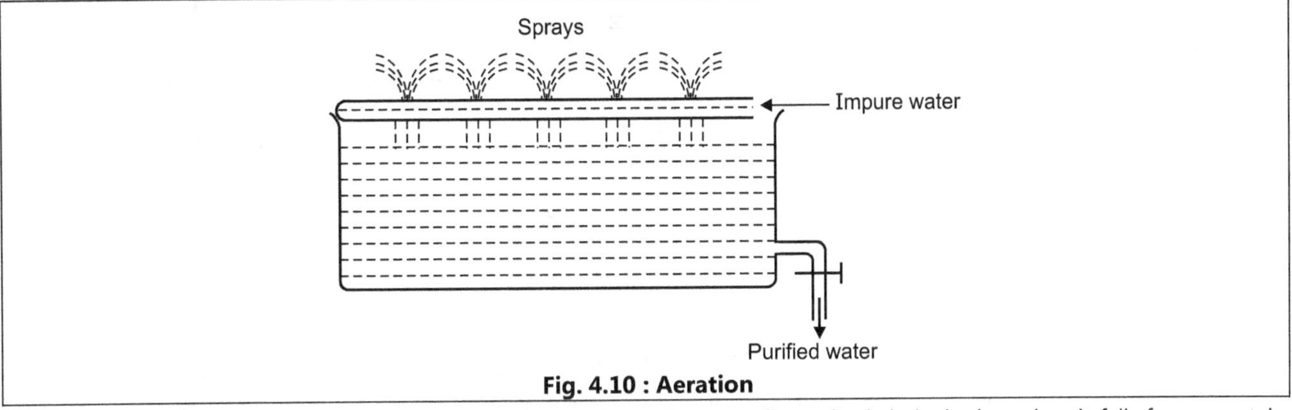

Fig. 4.10 : Aeration

- Natural aeration takes place in streams and rivers when the water flows slowly in its bed or when it falls from a certain height.

(v) Ultra-violet rays :

- The invisible ultra-violet rays are very effective in killing all types of bacteria.
- This method is widely used for the disinfection of swimming pool water, because it does not require any chemical to be mixed with water.
- In this method, ultra-violet rays (from mercury lamps enclosed in a quartz globe) are simply focussed on flowing water.
- However, this method is very costly and cannot be used for municipal supply water.

SOLVED EXAMPLES

Example 4.1 :

A water sample was found to contain following salts :

$Ca(HCO_3)_2$ = 40.5 mg/lit, $MgCl_2$ = 23.75 mg/lit, CO_2 = 3 mg/lit,

$MgCO_3$ = 21 mg/lit, $CaCl_2$ = 55.5 mg/lit, SiO_2 = 6 mg/lit.

Calculate the carbonate and non-carbonate hardness of the water sample.

Solution :

Step I : Conversion of the quantities of all the chemicals in terms of $CaCO_3$ equivalents in ppm.

	Salt/Chemical	Quantity in mg/lit	Mol. wt.	Type of hardness it causes in water	$CaCO_3$ equivalent in ppm
1.	$Ca(HCO_3)_2$	40.5	162	Carbonate	$40.5 \times \dfrac{100}{162} = \mathbf{25}$
2.	$MgCl_2$	23.75	95	Non-carbonate	$23.75 \times \dfrac{100}{95} = \mathbf{25}$
3.	$MgCO_3$	21	84	Carbonate	$21 \times \dfrac{100}{84} = \mathbf{25}$
4.	SiO_2	6	60	Does not contribute to hardness	–
5.	CO_2	3	44	–	–
6.	$CaCl_2$	55.5	111	Non-carbonate	$55.5 \times \dfrac{100}{111} = \mathbf{50}$

Step II : Calculation of carbonate or temporary hardness in ppm of $CaCO_3$

$$= [CaCO_3 \text{ equivalent in ppm of } Ca(HCO_3)_2 + MgCO_3]$$
$$= [25 + 25]$$
$$= \boxed{50 \text{ ppm}}$$

Step III : Non-carbonate hardness (permanent)

$$= [CaCO_3 \text{ equivalent in ppm of } MgCl_2 + CaCl_2]$$
$$= [25 + 50] = \boxed{75 \text{ ppm}}$$

Ans. Carbonate hardness = 50 ppm, Non-carbonate hardness = 75 ppm

---

Example 4.2 :

Calculate total, permanent and temporary hardness in ppm of $CaCO_3$ equivalents for the sample of water containing following salts :

$Ca(HCO_3)_2$ = 17.5 mg/lit, $Mg(HCO_3)_2$ = 14.6 mg/lit, $MgCl_2$ = 9.5 mg/lit,
$MgSO_4$ = 102.0 mg/lit, $MgCO_3$ = 8.4 mg/lit, $CaCl_2$ = 5.5 mg/lit,
$NaCl$ = 35 mg/lit.

Solution :

Conversion of all quantities of various chemicals in terms of $CaCO_3$ equivalents in ppm.

	Impurity	Quantity in mg/lit	Mol. wt.	Type of hardness	$CaCO_3$ equivalent, ppm
1.	$Ca(HCO_3)_2$	17.5	162	Carbonate	$17.5 \times \dfrac{100}{162} = 10.8$
2.	$Mg(HCO_3)_2$	14.6	146	Carbonate	$14.6 \times \dfrac{100}{146} = 10$
3.	$MgCl_2$	9.5	95	Non-carbonate	$9.5 \times \dfrac{100}{95} = 10$
4.	$MgSO_4$	12.0	120	Non-carbonate	$12 \times \dfrac{100}{120} = 10$
5.	$MgCO_3$	8.4	84	Carbonate	$8.4 \times \dfrac{100}{84} = 10$
6.	$CaCl_2$	5.5	111	Non-carbonate	$5.5 \times \dfrac{100}{111} = 5$
7.	$NaCl$	35	58.5	Does not cause hardness	

Calculation of carbonate (temporary hardness)

$$= [CaCO_3 \text{ equivalent in ppm of } Ca(HCO_3)_2 + Mg(HCO_3)_2 + MgCO_3]$$

$$= [10.8 + 10 + 10]$$

$$= \boxed{30.8 \text{ ppm}}$$

Calculation of non-carbonate permanent hardness

$$= [CaCO_3 \text{ equivalent in ppm of } MgCl_2 = MgSO_4 + CaCl_2]$$

$$= [10 + 10 + 5]$$

$$= \boxed{25 \text{ ppm}}$$

Example 4.3 :

25 ml of CaCl₂ solution (strength 250 mg CaCO₃ per 200 ml) required 35 ml EDTA solution. Same EDTA solution was used to titrate 25 ml of unknown hard water which consumed 30 ml of EDTA solution. Calculate the hardness of water sample.

Solution :

Strength of $CaCl_2$ solution is given as 250 mg of $CaCO_3$ per 200 ml.

That means 25 ml of $CaCl_2$ contains $= \left[\dfrac{25 \times 250}{200}\right]$ mg $CaCO_3$ = 31.25 mg $CaCO_3$

Now, 25 ml $CaCl_2$ solution requires 35 ml EDTA solution $= 31.25$ mg $CaCO_3$

∴ 1 ml EDTA solution $= \dfrac{31.25}{35}$ mg $CaCO_3$

Now,

25 ml hard water requires 30 ml EDTA solution $= \left[30 \times \dfrac{31.25}{35}\right]$ mg of $CaCO_3$

$= 26.8$ mg of $CaCO_3$

∴ 1000 ml hard water $= \left[\dfrac{26.8 \times 1000}{25}\right]$ mg of $CaCO_3$ equivalent hardness

Hardness of water $= 1072$ mg of $CaCO_3$

$= \boxed{1072 \text{ mg/lit or ppm}}$

Ans.   Hardness of water = 1072 ppm

Example 4.4 :

50 ml of standard hard water (1 mg CaCO₃/ml) required 35 ml EDTA solution while hard water sample required 20 ml EDTA solution. After boiling the requirement of EDTA was 12 ml. Calculate the total permanent and temporary hardness of water.

Solution :

Strength of standard hard water is 1 mg per ml $CaCO_3$.

i.e.   50 ml standard hard water $\equiv 50$ mg $CaCO_3$

50 ml standard hard water $\equiv 35$ ml EDTA solution

That means 35 ml EDTA solution $\equiv 50$ mg of $CaCO_3$

∴ 1 ml EDTA solution $= \dfrac{50}{35}$ mg of $CaCO_3$

Now,   50 ml unknown hard water $\equiv 20$ ml EDTA

$= \left[20 \times \dfrac{50}{35}\right]$ mg of $CaCO_3$

Thus total hardness per 50 ml $= \left[20 \times \dfrac{50}{35}\right]$ mg of $CaCO_3$

$$\therefore \quad \text{Total hardness per litre} \equiv \left[\frac{20}{35}\right] \times 1000$$

$$\equiv \boxed{571.42 \text{ ppm}}$$

Now,

$$\text{After boiling, 50 ml water sample} \equiv 12 \text{ ml EDTA}$$

Thus,

$$\text{Permanent hardness per 50 ml} = \left[12 \times \frac{50}{35}\right] \text{mg of } CaCO_3$$

$$= 17.14 \text{ mg of } CaCO_3$$

$$\therefore \quad \text{Permanent hardness per litre} = \left[17.14 \times \frac{1000}{50}\right] \text{mg of } CaCO_3$$

$$= \boxed{342.86 \text{ ppm}}$$

$$\text{Temporary hardness of water} = \text{Total hardness} - \text{Permanent hardness}$$

$$= (571.42 - 342.86) \text{ ppm}$$

$$= \boxed{228.56 \text{ ppm}}$$

Ans.
$$\text{Total hardness} = 571.42 \text{ ppm}$$
$$\text{Permanent hardness} = 342.86 \text{ ppm}$$
$$\text{Temporary hardness} = 228.56 \text{ ppm}$$

Example 4.5 :

50 ml of a standard hard water containing 1 mg of pure $CaCO_3$ per ml consumed 20 ml of EDTA. 50 ml of a water sample consumed 25 ml of same EDTA solution using Eriochrome black-T indicator. Calculate the total hardness of water sample in ppm.

Solution :

Here
$$50 \text{ ml of standard hard water} \Rightarrow 20 \text{ ml of EDTA}$$
$$50 \times 1 \text{ mg of } CaCO_3 \Rightarrow 20 \text{ ml of EDTA}$$
$$\therefore \quad 1 \text{ ml of EDTA} \Rightarrow 2.5 \text{ mg of } CaCO_3$$

Now, $\quad 50 \text{ ml of standard hard water} = 25 \text{ ml of EDTA}$

i.e. $\quad 25 \times 2.5 = 62.5 \text{ mg } CaCO_3$

$$\therefore \quad \text{Total hardness present per litre} = 62.5 \times \frac{1000}{50} \text{ mg/litre}$$

$$= 62.5 \times 20 \text{ mg/litre}$$

$$= 1250 \text{ mg/litre}$$

$$= \boxed{1250 \text{ ppm}}$$

Ans. $\quad \text{Total hardness} = 1250 \text{ ppm}$

Example 4.6 :

50 ml of standard hard water (1.2 g $CaCO_3$/lit) requires 32 ml of EDTA solution. 100 ml of water sample consumes 14 ml EDTA solution. 100 ml of boiled and filtered water sample consumes 8.5 ml EDTA solution. Calculate temporary hardness of this sample from above experimental data.

Solution :

Given data :

$$\text{Quantity of S.H.W. (1.2 g } CaCO_3/\text{lit)} = 50 \text{ ml}$$
$$\text{Quantity of EDTA consumed by 50 ml S.H.W.} = 32 \text{ ml}$$
$$\text{Quantity of hard water sample} = 100 \text{ ml}$$
$$\text{Quantity of EDTA consumed} = 14 \text{ ml}$$
$$\text{Quantity of EDTA consumed after boiling} = 8.5 \text{ ml}$$
$$\text{Hardness} = ?$$

Standardization of EDTA :

Standard hard water has 1.2 g i.e. $1.2 \times 1000 = 120$ mg of $CaCO_3$ equivalent H per lit.

120 mg/lit

1 ml – ?

$$\frac{120}{1000} = 0.12 \text{ mg/ml } CaCO_3 \text{ equivalent H}$$

∴   50 ml SHW $= 50 \times 1.2$ mg $CaCO_3 \equiv 60$ mg $CaCO_3$

Since   32 ml EDTA $= 50$ ml SHW $\equiv 60$ mg $CaCO_3$

$$1 \text{ ml EDTA} = \frac{60}{32} = 1.5 \text{ mg of } CaCO_3 \text{ equivalent H}$$

Calculation of total hardness :

100 ml hard water sample $= 14$ ml EDTA

∴ 14×1.5 mg of HW per 100 ml of HW $= 0.21$ mg of $CaCO_3$

∴   Per litre $= 210$ mg of $CaCO_3$

Calculation of permanent hardness :

Since   100 ml of boiled water $\equiv 8.5$ ml of EDTA

$= 8.5 \times 1.5$ mg of $CaCO_3$ in 100 ml

$= 0.1275$ mg of $CaCO_3$

Hence, $\boxed{\text{per litre} = 127.5 \text{ mg of } CaCO_3}$

Example 4.7 :

Calculate temporary hardness and total hardness of a sample of water containing $Ca(HCO_3)_2 = 16.2$ mg/lit,

$Mg(HCO_3)_2 = 7.3$ mg/lit, $CaSO_4 = 13.6$ mg/lit and $MgCl_2 = 9.5$ mg/lit.

Solution :

Conversion into $CaCO_3$ equivalents

Salt	Quantity	Multiplication factor	In terms of $CaCO_3$
$Ca(HCO_3)_2$	16.2 mg/lit	$\dfrac{100}{162}$	10 mg/lit
$Mg(HCO_3)_2$	7.3 mg/lit	$\dfrac{100}{146}$	5 mg/lit
$CaSO_4$	13.6 mg/lit	$\dfrac{100}{136}$	10 mg/lit
$MgCl_2$	9.5 mg/lit	$\dfrac{100}{95}$	10 mg/lit

Temporary hardness $= 10 + 5 = \boxed{15 \text{ mg/lit or ppm}}$

Permanent hardness $= 10 + 10 = \boxed{20 \text{ mg/lit or ppm}}$

Total hardness $= 15 + 20 = \boxed{35 \text{ mg/lit or ppm}}$

Example 4.8 :

Calculate the carbonate and non-carbonate hardness of a sample of water containing $MgCl_2$ = 9.5 ppm, $MgSO_4$ = 48 ppm, $Ca(HCO_3)_2$ = 16.2 ppm, KCl = 12 ppm, $Mg(HCO_3)_2$ = 14.6 ppm.

Solution :

Conversion into $CaCO_3$ equivalents

Salt	Quantity	Multiplication factor	$CaCO_3$ equivalent in ppm	Type of hardness
$Ca(HCO_3)_2$	16.2 ppm	$16.2 \times \dfrac{100}{162}$	10 ppm	Carbonate
$Mg(HCO_3)_2$	14.6 ppm	$14.6 \times \dfrac{100}{146}$	10 ppm	Carbonate
$MgCl_2$	9.5 ppm	$9.5 \times \dfrac{100}{95}$	10 ppm	Non-carbonate
$MgSO_4$	48 ppm	$48 \times \dfrac{100}{120}$	40 ppm	Non-carbonate
KCl	12 ppm	Does not contribute to hardness		

$$\text{Carbonate hardness} = [Ca(HCO_3)_2 + Mg(HCO_3)_2]$$
$$= [10 + 10] = \boxed{20 \text{ ppm}}$$
$$\text{Non-carbonate hardness} = [MgCl_2 + MgSO_4] = [10 + 40] = \boxed{50 \text{ ppm}}$$

Ans.
$$\text{Carbonate hardness} = 20 \text{ ppm}$$
$$\text{Non-carbonate hardness} = 50 \text{ ppm}$$

PRACTICE QUESTIONS

1. What is soft water ? What is hard water ?
2. What is hardness of water ?
3. Which principle is applied to remove hardness of water ?
4. Why hard water is unfit to use in boilers ?
5. Define : boiler corrosion, caustic embrittlement, priming and foaming, scale and sludge.
6. Define temporary hard water and permanent hard water.
7. List any two limitations of permutit or zeolite process.
8. What is lime soda process ? Name the chemicals used.
9. List two advantages of hot lime soda over cold lime soda.
10. What is permutit or zeolite ?
11. Write principle of ion-exchange process.
12. Which water is called demineralized water ?
13. What is demineralization ?
14. Define sedimentation, coagulation, filtration and sterilization.
15. Name the methods of sterilization.
16. Write principle of chlorination.
17. What are the characteristics of potable water ?
18. A sample of permanent hard water contains microorganisms. Name one method to remove microorganisms.
19. How cation exchange resins can be regenerated ?
20. Explain the impurities of water.
21. What is degree of hardness ?
22. Explain formation of scale and sludge formation in boilers.
23. What are the disadvantages of scale and sludge formation in boilers ?
24. Explain Clark's method of water softening.

❑❑❑

CORROSION

5.1 DEFINITION OF CORROSION (Nov. 18)

- Majority of the industrially important metals occur in nature in the form of their oxides, carbonates, sulphates, chlorides, silicates etc. These are reduced to their elementary form by metallurgical processes.
- The metals are then put to service. The exposed surface of almost every metal begins to decay more or less rapidly when it comes in contact with gaseous or liquid medium surrounding it. Thus, destruction of a metal starts.
- The decay of a metal is due to chemical or electrochemical reaction between the metal and the gases (present in air) or liquids and solutions.
- Thus, *"any process of chemical or electrochemical decay or destruction of a metal due to the action of surrounding medium is called as corrosion"*.
- Some examples of corrosion are : (1) rusting of iron, i.e. converting iron into its oxide (Fe_2O_3), (2) formation of a green film of basic carbonate on the surface of copper when exposed to moist air containing carbon dioxide gas.

5.1.1 Magnitude of Corrosion Problems

Corrosion is a slow process but the losses resulting from it are uneconomical due to the following reasons :

- The loss or damage resulting from corrosion cannot be measured by the cost of the metal alone, but also the high costs of its fabrication into machine tools or structure and the extraction of metals from their ores.
- Besides, the corrosion also causes a considerable reduction in the life of machinery, structure etc.
- The consequent defects in the machinery may cause casualities and injuries to workers.
- Moreover, repairing of damaged parts of machinery is costly.
- The other dangerous effects may be that of weakening of metallic structures such as bridges, boilers etc., and introduction of metallic poisons into water and food stuffs.
- Thus, metallic corrosion causes the above said multiple losses in one or the other way to the nation. Corrosion for engineers is a serious problem whether they are in design or in process operations.
- Every year losses due to corrosion of metals amount to crores of rupees.
- In India it is about 200 crore rupees. Therefore, now-a-days the scientists all over the world are concentrating themselves to understand the mechanism of corrosion and to develop various methods to protect the metal from corrosion.

5.2 TYPES OF CORROSION (Nov. 18)

- Corrosion occurs in many ways depending upon the attack of the metal, by the surrounding medium. There are two types of corrosion :
    1. Dry corrosion or direct chemical corrosion or atmospheric corrosion.
    2. Wet corrosion or electro-chemical corrosion or immersed corrosion.

5.2.1 Dry Corrosion or Direct Chemical Corrosion

- The corrosion occurs when metals come in contact directly with the atmospheric gases like O_2, CO_2 and moisture etc.
- The rate of corrosion is faster or quick in the humid and the polluted atmosphere near the industrial areas, as this atmosphere contains some corrosive gases like Cl_2, H_2S, SO_2 etc.
- The surface of metals are thus directly attacked by these gases present in the surrounding medium and gets coated with their corresponding compounds like oxides, sulphides, basic carbonates etc.

- Such a type of corrosion which is brought about by the atmospheric conditions is called *'dry corrosion' or direct chemical corrosion*. There are two main types of dry corrosion viz :

  1. Corrosion due to oxygen or oxidation corrosion.

  2. Corrosion due to other gases.

Oxidation corrosion :

- It has been found that the oxygen present in the medium is mainly responsible for corrosion of metal to a great extent.

- It is brought about by the direct action of oxygen at low or high temperatures on metals, usually, in the absence of moisture.

- The metallic surfaces when exposed to air undergo oxidation and the process is represented by the general equation

$$2M + O_2 \longrightarrow 2MO$$
$$\text{metallic oxide}$$

where M stands for the metal.

Mechanism of oxidation corrosion :

- Oxidation corrosion is brought about by direct action of oxygen on metals by forming oxide film.

- The mechanism of oxide film formation can be represented by the following reactions

$$M \longrightarrow \underset{\text{metal ion}}{M^{2+}} + 2e^- \qquad \text{(Loss of electron)}$$

$$O + 2e^- \longrightarrow O^{2-} \qquad \text{(Gain of electron)}$$

Net reaction, $\quad M + O \longrightarrow M^{2+} + O^{2-} \rightarrow MO \quad$ (Metal oxide)

- In above reactions, the electrons are transferred from metal atom to oxygen. Metal looses electrons while oxygen accepts electrons forming their respective ions. These two types of ions combine together to form metal oxide layer.

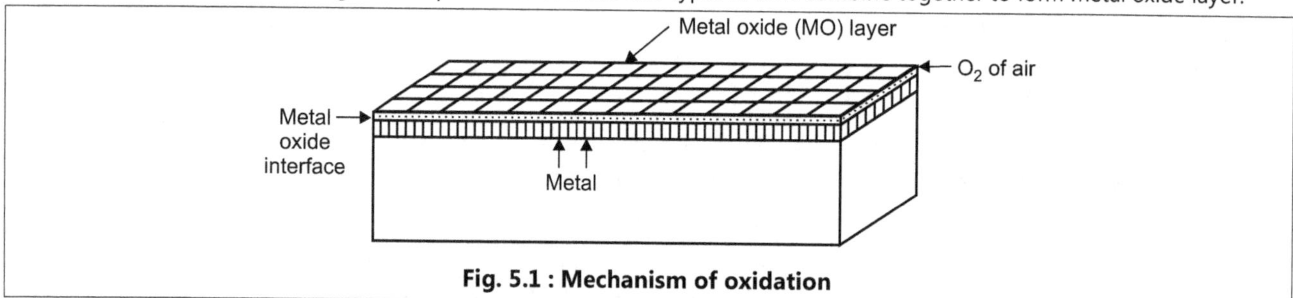

Fig. 5.1 : Mechanism of oxidation

- Nature of the oxide film formed on the surface of metal plays an important part in oxidation corrosion process. The film so produced can be classified into three categories : 1. Stable, 2. Unstable, 3. Volatile.

(i) Stable film :

Stable film (consists fine grains in its structure) is tightly adhering and impervious in nature.

- Such a film behaves as a protective coating in nature due to which after some time corrosion gets decreased to a great extent, if the film is not removed completely. This depends on the film being porous or non-porous.

(a) Porous (Non-protective) :

- This type of film is observed on the surfaces of alkali metals (like Li, K, Na etc.) and alkaline earth metals (like Ca, Sr, Mg).

- In these cases, volume of oxide film formed is insufficient to cover completely the comparatively larger surface area of the metal. Consequently, the oxide layer faces stresses and strains, thereby developing cracks and pores in its structure.

- Hence, porous oxide layer permits free access of oxygen to the fresh metal surface below (through cracks and pores) and thus corrosion continues.

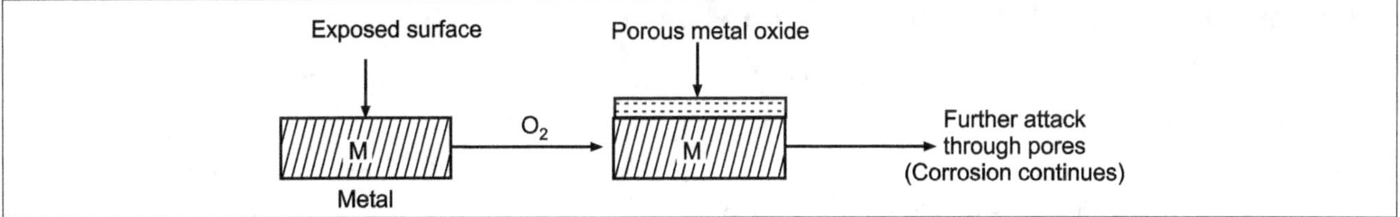

(b) Non-porous (Protective) :

- The metal like aluminium forms oxide (Al_2O_3) whose volume is greater than the volume of metal (Al).

- This oxide film is extremely adherent and non-porous (protective). Due to absence of pores or cracks in the oxide film, it forms a barrier for further action and therefore, the rate of oxidation of metal rapidly decreases.

- The oxide film formed on the surface of metals like aluminium, tin, copper, lead are stable (protective) and hence prevent corrosion.

- Therefore, though the tendency of aluminium to oxidation is much greater than iron, aluminium gets protected from corrosion due to formation of non-porous, tightly adhering, protective oxide film on its surface which does not permit corrosion to occur.

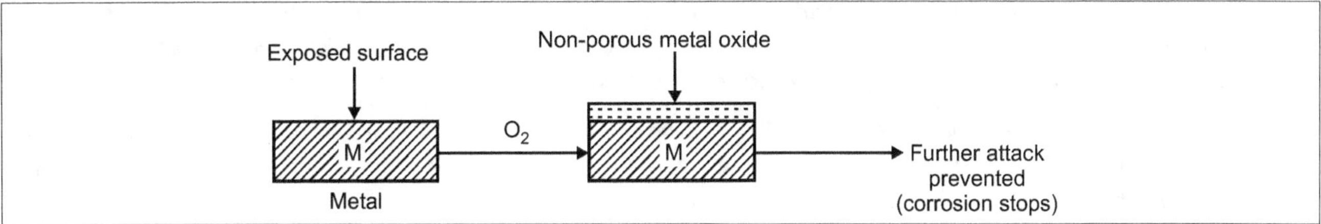

(ii) Unstable film :

- When the oxide film is unstable, it decomposes back into the metal and oxygen, as soon as it is formed.

$$2MO \longrightarrow 2M + O_2 \uparrow$$

(metal oxide) (metal) (oxygen)

- Therefore corrosion is not possible in case of noble metals such as silver, gold, platinum etc.

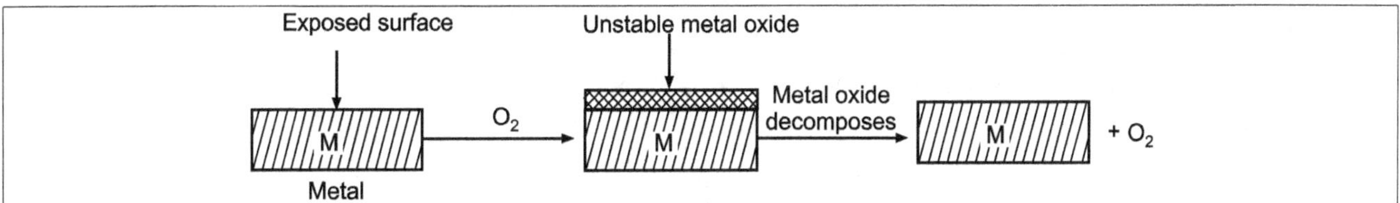

(iii) Volatile film :

- When the oxide film is volatile, it vaporises as soon as it is formed on the metal surface. Therefore the fresh metal surface is exposed continuously to the atmosphere.

- This causes rapid and continuous corrosion. For example, molybdenum oxide (MoO_3) and stannic chloride ($SnCl_4$) in the presence of dry chlorine are volatile films.

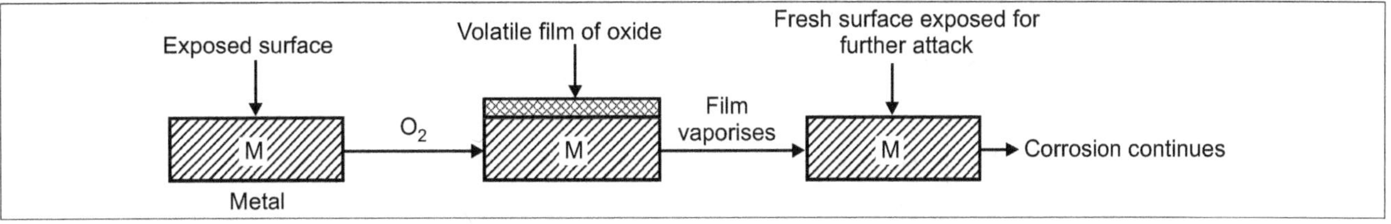

5.2.2 Wet Corrosion or Electro-chemical Corrosion or Immersed Corrosion

- This type of corrosion occurs at a solid-liquid interface when the metals are in contact with moist air or any liquid medium.

- If two dissimilar metals are dipped in a solution, the solution acts as a conducting medium between them. One of the two metals acts as the anode and the other as a cathode.

- The anode gives off ions in the solution, so anodic metal is destroyed by either dissolving or assuming combined state (such as oxide etc.), *hence anode gets corroded.*

- On the other hand, cathode receives the ions and forms a protective coating, hence it is not corroded.

At anode : $M \longrightarrow M^{n+} + ne^-$ (oxidation)

　　　　　metal　　　　　metal ion

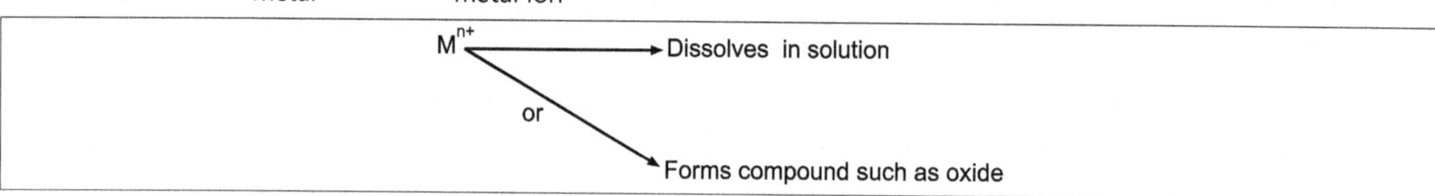

- *Thus, "the corrosion which is brought about through ionic reactions in the presence of moisture or solution as a conducting medium when two dissimilar metals are in contact with each other is called as wet corrosion or electrochemical corrosion".* It is also known as immersed corrosion as it occurs in metals when they are immersed or dipped in the same solution.

- The wet corrosion takes place by the formation of several types of electrolytic cells, mainly galvanic or concentration cells of different magnitude depending upon the nature of the metal and the surrounding medium, which ultimately promote the corrosion of the metal.

  (a) "Galvanic cells" are set up when two dissimilar metals come in contact with each other and are surrounded by moist air or liquid medium.

  (b) "Concentration cells" are set up when a single metal is surrounded by moist air with difference of ion concentration in solution of that of oxygen.

- The action of such types of cells promoting the corrosion of metals is described below.

5.2.2.1 Galvanic Cell Corrosion

- When two dissimilar metals are electrically connected and dipped in an electrolyte solution, a galvanic cell is formed.

- Working of galvanic cell can be explained by considering a Daniel cell, which consists of zinc and copper plates dipped in solutions having Zn^{++} and Cu^{++} ions respectively.

- Two solutions are separated by a porous partition and Cu and Zn plates are connected by a metallic conductor.

- The e.m.f. developed is due to two separate reactions taking place at the electrodes. Zn metal being higher in the electrochemical series acts as a anode and goes into the solution as Zn^{++} ions with the liberation of two electrons. This is the oxidation process.

$$Zn \text{ (solid)} \longrightarrow Zn^{++} \text{ (aq)} + 2e^- \text{ (oxidation i.e. de-electronation)}$$

- The liberated electrons move along metallic conductor, and are accepted by copper which acts as a cathode and it undergoes the process of reduction.

$$Cu^{2+} \text{ (aq)} + 2e^- \longrightarrow Cu \text{ (solid)} \text{ (reduction i.e. electronation)}$$

- Thus, zinc goes into the solution (gets dissolved) and corroded, while copper is deposited at copper cathode and gets protected.

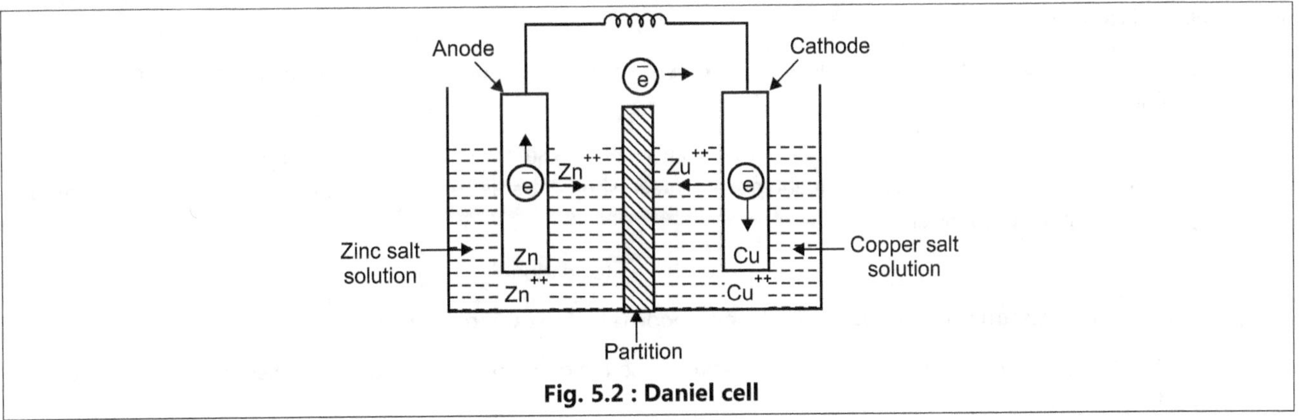

Fig. 5.2 : Daniel cell

- The nature of the corrosive environment decides the type of cathodic reactions.

- In acidic solution, the corrosion occurs by the hydrogen evolution process while in neutral or slightly alkaline solution, oxygen absorption occurs.

- The electron current flows from the anodic metal (Zn) to the cathodic metal (Cu). Thus, it is evident that the corrosion occurs at the anodic metal while the cathodic metal is protected from the attack.

- Thus, the following points are necessary to undergo electrochemical corrosion :

    (i) Formation of anodic and cathodic areas.

    (ii) Electrical contact between anode and cathode for movement of electrons.

    (iii) Presence of conducting medium.

Examples :

    (i) Rusting of fencing wire under joints.

    (ii) Corrosion at revietted joints.

    (iii) Steel pipe connected to copper plumbing.

    (iv) Lead-antimony solder around the copper wire.

    (v) Steel screws in marine brass-hardware.

    (vi) Copper sheets joined by iron nails.

In above examples, more than one metal is used in the structure. The anodic metal will corrode while others will be protected.

Mechanism of wet corrosion by galvanic cell action :

- Wet corrosion or electrochemical corrosion involves flow of electron current between anodic and cathodic areas.

- The reaction at anode is dissolution of metal as its metallic ion with the liberation of free electrons.

    At anode : $M \longrightarrow M^{n+} + ne^-$ (oxidation)

- The reaction at cathode is consumption of electrons with

    (a) Evolution of hydrogen gas or

    (b) Absorption of O_2, depending upon the nature of environment.

Hydrogen evolution mechanism :

- This type of corrosion occurs when metals are exposed to acidic environments like industrial waste, solutions of non-oxidising acids (like HCl).

- For example, a steel tank containing acidic industrial waste and a small piece of copper scrap is in contact with the steel tank. In this case, the portion of the steel tank in contact with copper is corroded most with the evolution of hydrogen gas. The chemical reactions are

$$Fe \longrightarrow Fe^{++} + 2e^- \text{ (oxidation)}$$

- These liberated electrons flow through metal, from anode (steel) to cathode (copper).

- The hydrogen ions (H^+) from acid accept the electrons and are deposited at cathode from where evolution of hydrogen gas takes place.

$$2H^+ + 2e^- \longrightarrow H_2 \uparrow \text{ (reduction)}$$

Thus, overall reaction is :

$$Fe + 2H^+ \longrightarrow Fe^{++} + H_2 \uparrow$$

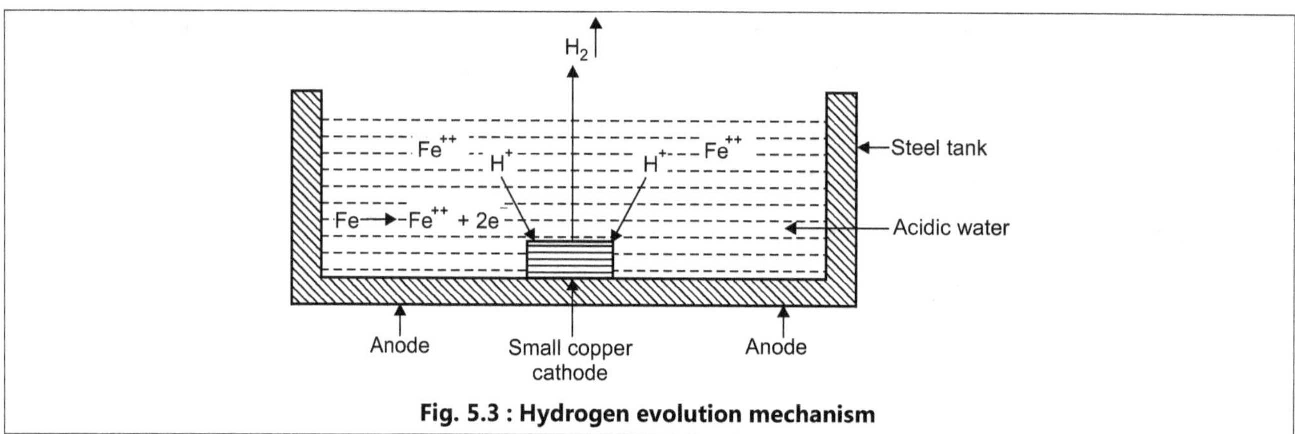

Fig. 5.3 : Hydrogen evolution mechanism

- It may be pointed here that in this type of corrosion (i) all metals above hydrogen in the electrochemical series have a tendency to get dissolved in acidic solution with simultaneous evolution of hydrogen gas. (ii) The anodes are usually very large areas, whereas the cathodes are small areas.

5.2.2.2 Concentration Cell Corrosion

- This type of corrosion is due to electrochemical attack on the metal surface, exposed to an electrolyte of varying concentrations or of varying aeration.

- The most common type of concentration cell corrosion is differential aeration corrosion.

- It occurs when one part of metal is exposed to a different air concentration from the other part. This causes a difference in potential between differently aerated areas.

- It has been found experimentally that "poor oxygenated parts are anodic". Consequently, a differential aeration of metal causes a flow of current, called the differential current.

- Differential aeration accounts for the corrosion of metals, partially immersed in a solution, the waterline just below.

- Iron corrodes under drops of water (or salt solution) in the presence of atmospheric oxygen (oxygen absorption mechanism) is a common example of concentration cell corrosion.

- The surface of iron is usually coated with a thin film of iron oxide. However, if this iron oxide film develops some cracks, anodic areas are created on the surface, while the coated metal parts act as cathode.

- It follows that the anodic areas are small surface parts, while nearly rest of the surface of the metal forms large cathodes.

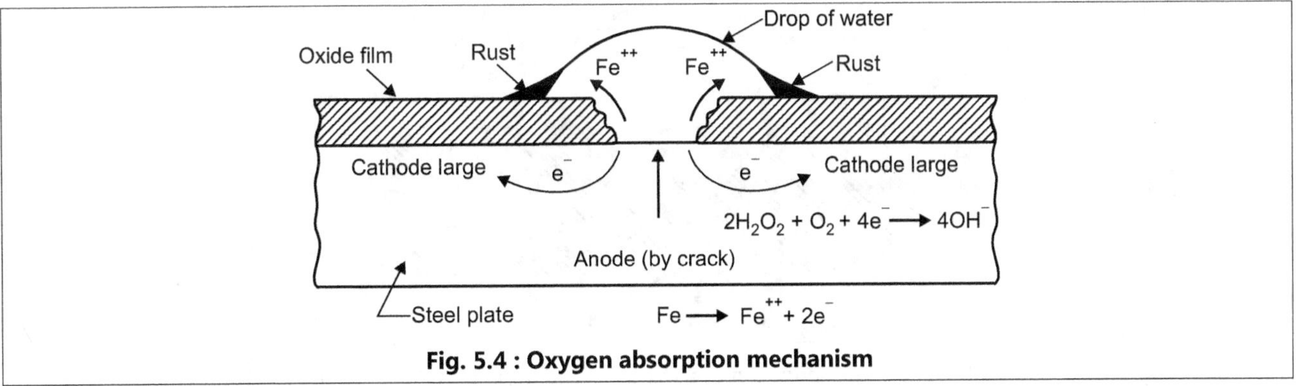

Fig. 5.4 : Oxygen absorption mechanism

- At the anodic areas of the metal, iron dissolves as Fe^{++} ion with the liberation of electrons.

$$Fe \longrightarrow Fe^{++} + 2e^-$$

- The liberated electrons flow from anodic to cathodic areas, through iron metal, where electrons are intercepted by the dissolved oxygen. These in presence of water drop form OH^- ions as follows :

$$2H_2O + O_2 + 4e^- \longrightarrow 4(OH^-)$$

- The Fe^{++} ions at anode and OH^- ions at cathode diffuse and when they meet, ferrous hydroxide is precipitated.

$$Fe^{++} + 2OH^- \longrightarrow Fe(OH)_2\downarrow$$

(i) If enough oxygen is present, ferrous hydroxide is easily oxidised to ferric hydroxide.

$$4Fe(OH)_2 + O_2 + 2H_2O \longrightarrow 4Fe(OH)_3$$

This product, called *'yellow rust'* actually corresponds to $Fe_2O_3.H_2O$.

(ii) If the *'supply of oxygen is limited'*, the corrosion product may be even black anhydrous magnetite, Fe_3O_4.

5.2.3 Pitting Corrosion

- Pitting corrosion is a localised accelerated attack at some places on the metal surface, resulting in the formation of pits, holes or cavities in the metal.

- Pitting corrosion occurs due to breakdown or cracking of the protective film on a metal at specific point.

- Breakdown of the protective film may be caused by various mechanical factors such as surface roughness, scratches or cut edges, local straining of metal because of non-uniform stresses, sliding under load or chemical attack over the surface of metal.

- As a result of pitting, small anodic and large cathodic areas are developed and this gives rise to corrosion current in correct environment.

- Pitting is neither uniform nor has a constant rate.

- Pitting occurs mostly in chloride solution containing oxygen or oxidising salt.

- Pitting is also caused by the presence of sand, dust, scale and other extraneous impurities present on the metal surface.

- Because of the different amounts of oxygen in contact with the metal, the small part underneath the impurity becomes the anodic areas and the surrounding area becomes cathodic areas.

- Intense corrosion takes place in the anodic areas underneath the impurity.

- The rate of corrosion will be increased after the formation of a small pit or cavity or hole in the beginning.

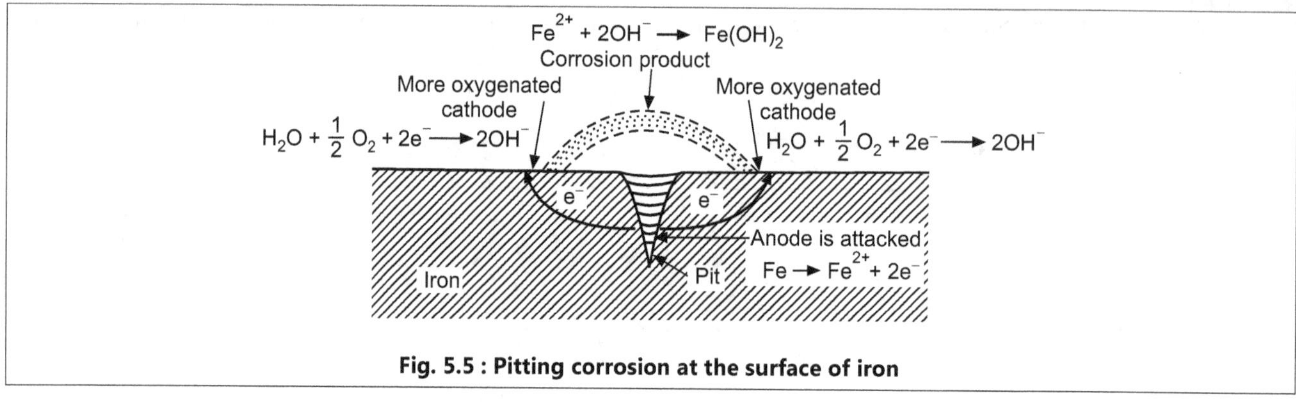

Fig. 5.5 : Pitting corrosion at the surface of iron

- Pitting can be reduced by reducing chloride ion concentration and in stainless steel, pitting can be reduced by adding 3-4% molybdenum.

5.2.4 Waterline Corrosion

- When water is stored in a steel tank, it is generally found that the maximum amount of corrosion takes place along a line just beneath the level of the water meniscus.

- The area above the waterline (highly oxygenated) acts as cathodic area and is completely unaffected by corrosion. However, little corrosion takes place when the water is relatively free from acidity.

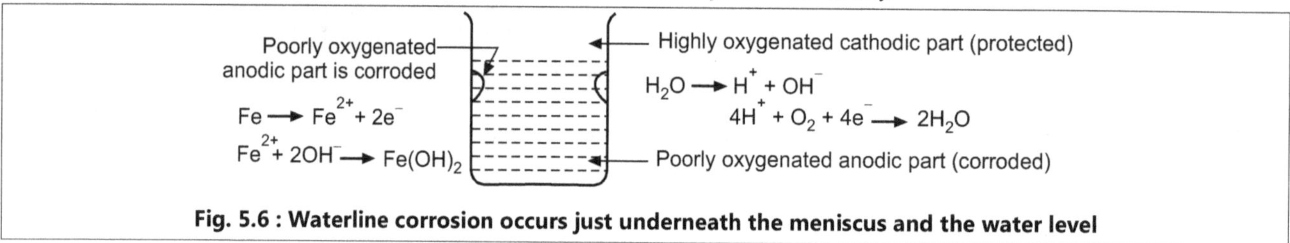

Fig. 5.6 : Waterline corrosion occurs just underneath the meniscus and the water level

- Waterline corrosion is also caused in marine ships and is accelerated by marine plants which are attached to the sides of ships.

- This type of corrosion is prevented to a great extent by painting the sides of ships by special antifouling paints.

5.2.5 Crevice Corrosion

- Crevice corrosion refers to the localized attack on a metal surface at, or immediately adjacent to, the gap or crevice between two joining surfaces.

- The gap or crevice can be formed between two metals or a metal and non-metallic material.

- Outside the gap or without the gap, both metals are resistant to corrosion.

- The damage caused by crevice is normally confined to one metal at localized area within or close to the joining surfaces.

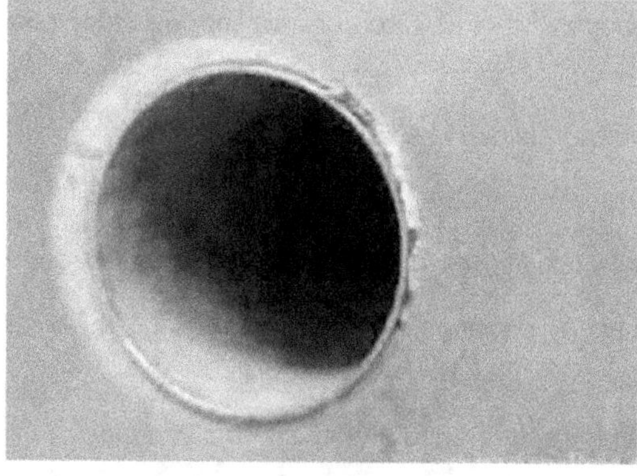

Fig. 5.7 : Crevice corrosion

- In Fig. 5.7, crevice corrosion is seen due to the presence of crevice (gap) between stainless steel tube and tube sheet of heat exchanger used in a sea water reverse osmosis (SWRO) desalination plant.

- Crevice corrosion is initiated by a difference in concentration of oxygen, which set-up an electrochemical concentration cell (differential aeration cell).

- Oxygen concentration is higher on outside the crevice (the cathode) comparatively inside the crevice (the anode). Hence anodic part undergoes fast corrosion.

- Once a crevice has formed, the propagation mechanism for crevice corrosion is the same as for pitting corrosion.

- Crevice corrosion can be prevented by

    (1)   Use of welded butt joints instead of riveted or bolted joints in new equipment.

    (2)   Eliminate crevices in existing lap joints by continuous welding or soldering.

    (3)   Avoid creating stagnant conditions and ensure complete drainage in vessels.

    (4)   Use solid, non-absorbant gaskets such as teflon.

    (5)   Use suitable alloys for increased resistance to crevice corrosion.

5.2.6 Galvanic Series

- It is an electrochemical series obtained by arranging metals in the order of their standard electrode potential (reduction electrode potential) arranged down in an increasing order.

- In this series, the metals with high negative potential (low reduction potential) are chemically reactive and with high positive potential (high reduction potential) are noble. Therefore, chemical reactivity or passivity of metals can be predicted from their position.

- A metal high in the series is more anodic and undergoes corrosion faster than the metal below it.

- For example, Li corrodes faster than Mg, Zn corrodes faster than Fe, Fe corrodes faster than Sn, Cu corrodes faster than Ag and so on.

Table 5.1: Standard electrode potential (reduction) at 25°C

Element	Electrode	Electrode Potential in volt E^o	Reference
Lithium	Li^+	-2.95	
Potassium	K^+	-2.92	
Calcium	Ca^{++}	-2.76	
Sodium	Na^+	-2.71	
Magnesium	Mg^{++}	-2.37	Anodic
Aluminium	Al^{+++}	-1.69	
Zinc	Zn^{++}	-0.76	
Chromium	Cr^{+++}	-0.71	
Iron	Fe^{++}	-0.44	
Nickel	Ni^{++}	-0.23	
Tin	Sn^{++}	-0.14	
Lead	Pb^{++}	-0.13	
Hydrogen	H^+	-0.00	Reference
Copper	Cu^{++}	$+0.34$	
Silver	Ag^+	$+0.80$	
Gold	Au^{+++}	$+1.50$	Cathodic

5.3 FACTORS AFFECTING THE RATE OF CORROSION   (Nov. 18)

The rate and extent of corrosion depend upon mainly the following :

Nature of the metal	Nature of the environment
1. Position in the galvanic series.	1. Temperature.
2. Purity of metal.	2. Humidity (moisture).
3. Physical state of the metal.	3. Effect of pH.
4. Nature of the oxide film.	4. Conductance of the medium.
5. Relative areas of anode and cathode.	5. Differential aeration.
6. Solubility of the corrosion product.	6. Presence of impurities.

(I) NATURE OF THE METAL :

Let us discuss first, how rate and extent of corrosion depend on nature of the metal.

(i) Position of metal in a galvanic series :

- A metal having higher position in a galvanic series, has more chemical reactivity and therefore, it gets attacked by gaseous corroding medium faster.

- However, if the film of the corrosion product is non-porous, then corrosion stops naturally.

- When two dissimilar metals are in contact in an aqueous medium, then the metal having higher position in a galvanic series, gets corroded.

- Greater the difference in their positions, faster is the rate of corrosion by the electrochemical mechanism.

(ii) Purity of the metal :

- Impurities present in a metal cause heterogeneity and form a large number of tiny galvanic cells when an aqueous medium comes in contact with such a metal.

- If the impurity metal is highly placed in a galvanic series then it acts as a anode and gets corroded to produce small depressions on the surface of the base metal.

- Thus, corrosion resistance of a metal can be improved by increasing its purity.

(iii) Physical state of the metal :

- The physical state of the metal means orientation of crystals, grain size, stress etc.

- The larger the grain size of the metal/alloy, the smaller will be its solubility and hence, lesser will be its corrosion. e.g. mild steel grains are smaller than cast iron grains, therefore, mild steel gets corroded faster.

- Areas under stress (bend, joints, rivets), tend to be anodic and corrosion takes place at these stressed areas.

- The grain size in a metal can be increased by hardening operation or by alloying with a suitable element.

(iv) Nature of the oxide film :

- Metals are having oxide film formed on their surface by corrosion due to oxygen.

- If the film is non-porous then the metal safeguards itself from further corrosion. But if the film is porous then the corrosion by the gas continues until total destruction of the metal.

(v) Relative areas of anode and cathode :

- The rate of corrosion of a metal is less when the area of the cathode is smaller.

- When cathodic area is smaller, the demands for electrons will be less and this results in decreased rate of dissolution of the metal at anodic regions.

In general, rate of corrosion of anodic region $\propto \dfrac{\text{Cathodic area}}{\text{Anodic area}}$.

- For example, a small steel pipe fitted in a large copper tank undergoes localized, rapid and severe corrosion.

(vi) Solubility of the corrosion product :

- Insoluble corrosion products function as a physical barrier, thereby, suppresses further corrosion.

- But if the corrosion product is soluble in the corroding medium, the corrosion of the metal proceeds faster.

- For example, Pb in H_2SO_4 medium forms $PbSO_4$, which is insoluble in the corroding medium (H_2SO_4), hence corrosion proceeds at a smaller rate.

(II) NATURE OF THE ENVIRONMENT :

(i) Temperature :

- The rate of atmospheric and electrochemical corrosion become faster at higher temperature. Gas molecules and metal atoms, both become more active in dry corrosion at higher temperature.

- The cell reactions also become faster during wet corrosion, at higher temperature.

(ii) Humidity :

- The greater the humidity, the greater is the rate and extent of corrosion.

- If the moisture percentage is high (in rainy season, fog, costal areas etc.) then it can act as aqueous conducting medium and causes setting up of cells. Thus high humidity brings about electrochemical corrosion.

- We know that generally the rate of electrochemical corrosion is faster than the rate of atmospheric corrosion.

  For example, atmospheric corrosion of Fe is slow in dry air as compared to moist air.

(iii) Effect of pH :

- Corrosion of those metals which are readily attacked by acids can be reduced by increasing the pH of the attacking environment.

- For example, corrosion of Zn can be minimized by increasing the pH to 11.

- In general, acidic media are more corrosive than alkaline and neutral media.

(iv) Conductance of the medium :

- A dry sandy soil has much less conductance and therefore the rate of electrochemical corrosion of the metal in contact with such soil is very less. While the clay, mineralised soils, sea water etc. have much higher conductance and cause fast corrosion of the metal by electrochemical reactions.

- The stray currents from power leakages damage the metal structures buried under the soils of higher conductivity.

(v) Differential aeration :

- This occurs when one part of the metal is exposed to a different air concentration than the other. It is found that poor or least oxygenated part becomes anodic to the remaining part of the metal and it is corroded.

- Thus, "corrosion occurs where oxygen access is least", this is known as differential aeration principle and was developed by Evans. The examples of corrosion due to differential aeration are :

  (1) When a pipeline passes through moist soil as well as dry soil, the part passing through moist soil, having restricted oxygen access becomes anodic, while the part passing through dry soil, having more access of air, becomes cathodic. This causes corrosion of pipe embedded in the moist soil.

  (2) Waterline corrosion is quite common in ships, water storage iron tanks etc. in which a portion of the metallic surface is under water (less oxygenated water), while the remaining portion is above the water (more oxygenated water). This results in the formation of a cell in which the anode is the lower portion, while the cathode is at the water level. Due to this corrosion occurs just below the water level.

(vi) Presence of impurities in the atmosphere :

- Corrosion of metals is more in areas near to the industry and sea. This is because corrosive gases like H_2S, SO_2, CO_2 and fumes of H_2SO_4 and HCl in industrial areas and NaCl of sea water lead to increase conductivity of the liquid layer in contact with the metal surface, thereby increases the rate of corrosion.

5.4 PROTECTION OF METAL FROM CORROSION (CORROSION CONTROL)

- The various protective measures include :

(1) Modification of the environment.

(2) Cathodic protection.

(3) Use of protective coatings (metal coating).

- Before protecting measures are adopted, the metal should be cleaned by degreasing and descaling.

- Degreasing and descaling can be done by chemical or mechanical methods.

**1.	Modification of the environment :**

- In this category, the metals are protected from corrosion either by removal of corrosion stimulants or by the use of inhibitors (i.e. the substances which effectively decrease the corrosion rate).

	These are briefly discussed below :

**(a)	Removal of corrosion stimulants :**

- To prevent corrosion due to oxygen, dissolved oxygen from water is removed by physical or chemical means. Either deaeration is done or reducing substances are added like N_2H_4, Na_2SO_3 etc.

- To prevent corrosion by acids, they are neutralized with lime.

- To prevent corrosion by salts, they are removed by using ion-exchange resins.

- To prevent corrosion by moisture (humidity), moisture from air is removed by dehumidification using silica gel.

**(b)	Use of corrosion inhibitors :**

- Inhibitors are organic or inorganic substances which when added to the environment are able to reduce the rate of corrosion.

- Inhibitors form a film which imposes a physical barrier between the metal and the medium.

- Quinoline, organic amines, cyanides, chromates are effective inhibitors.

- For example, potassium chromate when added to the medium stop the corrosion of iron and other metals by oxidising iron to ferric state and forming a protective film of insoluble oxide over the anode surface.

**2.	Cathodic protection :**

- The principle of this method is the metal to be protected is forced to behave as a cathode. There are two ways to do cathodic protection.

**(A)	Using sacrificial anode :**

- The metallic structure to be protected from corrosion is connected to the anodic metal (active) by an insulating wire.

- The more active metal like Zn, Al, Mg etc. acts as a anode and gets corroded, hence it is known as sacrificial anode. For the purpose of increasing electrical contact, the active metal is placed in back fill (Coal + NaCl).

- When the sacrificial metal is consumed completely, it is replaced by fresh piece.

- This method is applicable to protect buried pipelines, buried cables, hot water tank, ship hull etc.

	Mg or Zn rods are bolted along the sides of ship, hot water tank or inserted into boiler to prevent corrosion.

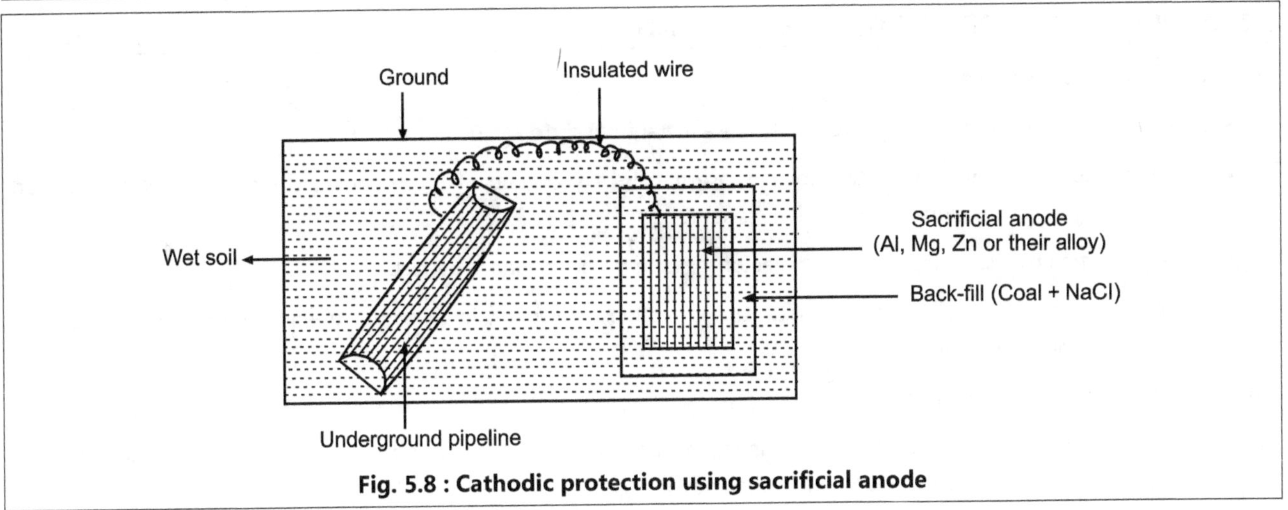

Fig. 5.8 : Cathodic protection using sacrificial anode

(B) Using impressed current :

- In this method, an impressed current is applied in opposite direction to nullify the corrosion current and convert the corroding metal from anode to cathode.

- The impressed current is derived from a D.C. source and given to insoluble anode like graphite, stainless steel or scrap iron, buried in soil. The negative terminal of D.C. is connected to the pipeline to be protected.

- The anode is kept in back-fill (composed of gypsum or coke breeze), to increase electrical contact with the surrounding soil.

- This type of cathodic protection is applicable to :

Open water box coolers, water tanks, buried water or gas pipeline, condensers, transmission line towers, marine piers etc.

This method is particularly useful where current requirements and resistivity are high. This is suited to large structures and long-term operation.

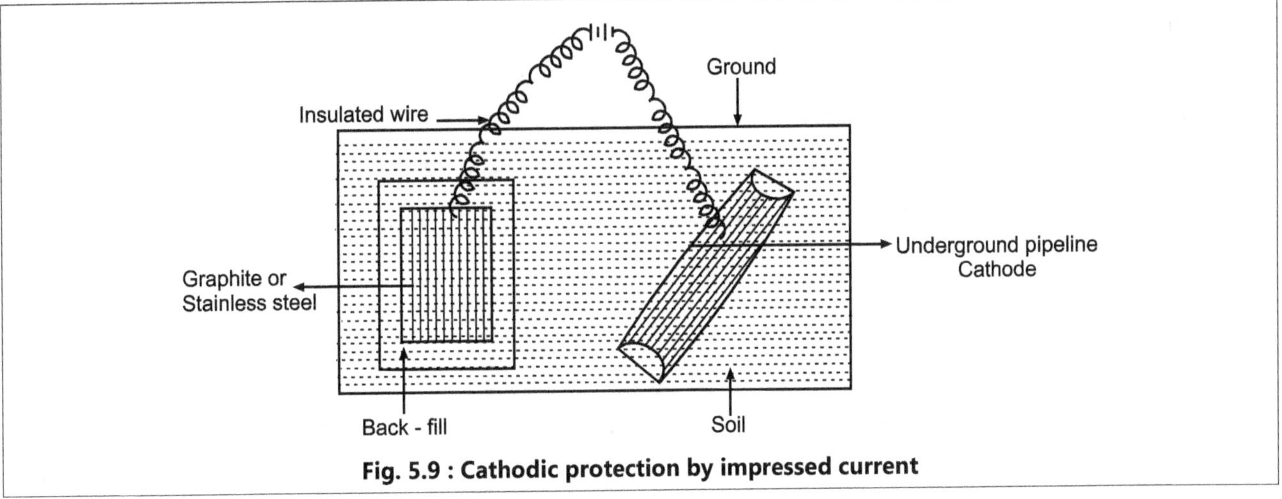

Fig. 5.9 : Cathodic protection by impressed current

3. Metal coating :

- Coating of metal is very useful for protecting the metal from corrosion. It can be classified as

(a) Coating of more active metals. (b) Coating of less active metals.

## 5.5 APPLICATIONS OF PROTECTIVE COATINGS	(Nov. 18)

(a) Coating of less active metals :

- Less active metals (noble metals) protect the base metal from corrosion due to their inactivity.

- They provide mechanical protection to the base metals. e.g. coating of tin in tinning is made to protect iron from corrosion because it is extremely corrosion resistant.

- The protection by less active metal is satisfactory as long as the coating is perfect.

- A break or crack in coating may facilitate the formation of electrolytic cells due to which, more active base metal undergoes corrosion rapidly.

(b) Coating of more active metals :

- Coating of more active metals like zinc, aluminium, cadmium protect the base metal from corrosion. In this case, base metal acts as a cathode and coating metal becomes an anode.

- So the two metals being in contact with the surrounding medium form a galvanic cell in which cathode remains unattacked, while the anode corrodes.

- Such a protection is also known as cathodic protection. For example, coating of zinc on iron. i.e. galvanising.

Methods of applying metal coating :

The protective metallic coating can be made by the following different methods :

1. Hot dipping : (i) Galvanising, (ii) Tinning.

2. Metal spraying.

3. Electroplating.

4. Metal cladding.

5. Cementation or diffusion coating : (a) Colorizing, (b) Chromizing, (c) Sherardizing.

1. Hot dipping :

- This method is based on the fact that the melting point of protective metal should be lower than the melting point of the base metal.

- Generally coating metals are zinc (melting point 419°C) and tin (melting point 232°C) and base metals are iron, steel and copper, which have relatively higher melting points.

- The process consists of dipping the base metal in a bath of molten coating metal, covered by a molten flux layer.

- The flux cleans the surface of base metal and prevents the oxidation of molten coating metal.

- Hot dipping is done by two methods :

  (a) Galvanising. (b) Tinning.

(a) Galvanising :

- Galvanising is the process of coating iron or steel sheet with a thin coat of zinc to prevent it from rusting.

- Galvanising is the example of cathodic protection, in which zinc acts as an anode and undergoes corrosion; while base metal acts as a cathode and gets protected.

- If protective zinc coating is broken, then iron too is protected because zinc is more electropositive than iron and therefore it does not allow iron to go into the solution.

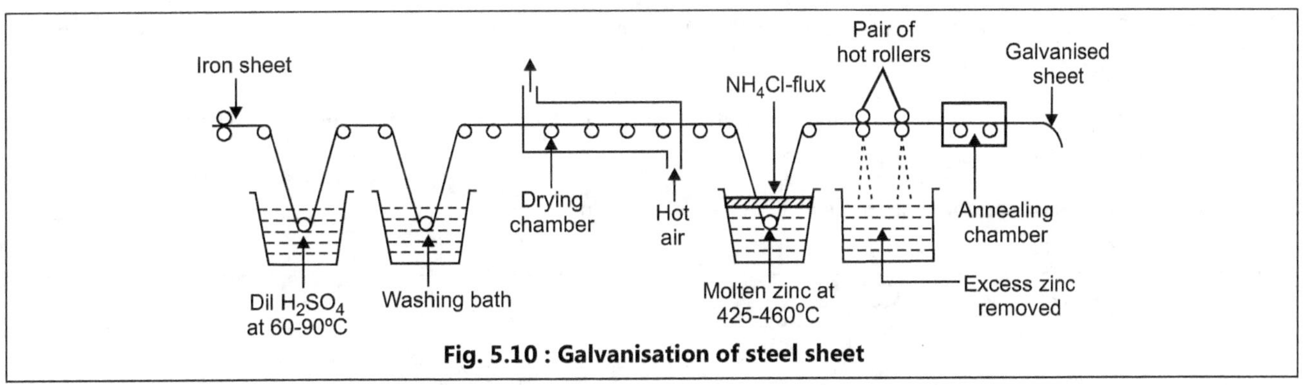

Fig. 5.10 : Galvanisation of steel sheet

- The iron or steel article to be galvanised is first cleaned with dilute sulphuric acid to remove any superficial oxide layer and impurities.

- It is then dipped in a bath of zinc ammonium chloride solution and then allowed to dry.

- The sheet is then dipped in another bath containing molten zinc at the temperature of 425°C to 460°C.

- The surface of the bath is kept covered with a flux of ammonium chloride to prevent oxide formation. When the sheet is taken out, it is found to have been coated with zinc.

- It is then passed through pairs of hot rollers to make the coating uniform, and to remove the superfluous (excess) metal, if any.

- Finally, it is annealed at a temperature of 650°C and then slowly cooled.

Applications of Galvanising process

- It is widely used for protecting iron exposed to the atmosphere as in the case of roofing sheets, fencing wires, pipes, bolts, screws, nails, buckets, tubs etc.

- Galvanised ware (zinc coated utensils) cannot be used for preparing and storing food stuffs, especially which are acidic in nature because zinc gets dissolved in dilute acids forming poisonous zinc compounds which will poison the contents.

Advantages of Galvanising :

- Low installation and maintenance cost.

- Long endurance.

- Smooth and high quality coating even on inconvenient and complicated surfaces.

- Automatic protection for damage areas.

- High resistance to mechanical wear and damage.

- Equally good coating on sharp edges and corners.

Disadvantages :

- Galvanised containers cannot be used for preparing and storing food stuffs.

(b) Tinning :

- It is the process of covering iron or steel sheets with a thin coat of tin to prevent it from corrosion.

- Tinning is very similar to galvanising, except that molten tin is used in place of zinc.

- Tin is less electropositive metal than iron and therefore, it is more resistant to chemical attack. Moreover, it is having low melting point (232°C).

- In case, if tin coating formed does not cover the surface completely or it is broken during use and leaves iron surface exposed, then more rapid corrosion of iron will take place than that of unprotected surface of iron.

- This is because iron stands above tin in the activity series and is therefore, more electropositive than tin. Thus, iron easily loses electrons and undergoes corrosion rapidly.

- Tin is extensively used for coating over mild steel plates required for food stuff industry as it is resistant to action of organic acid and water.

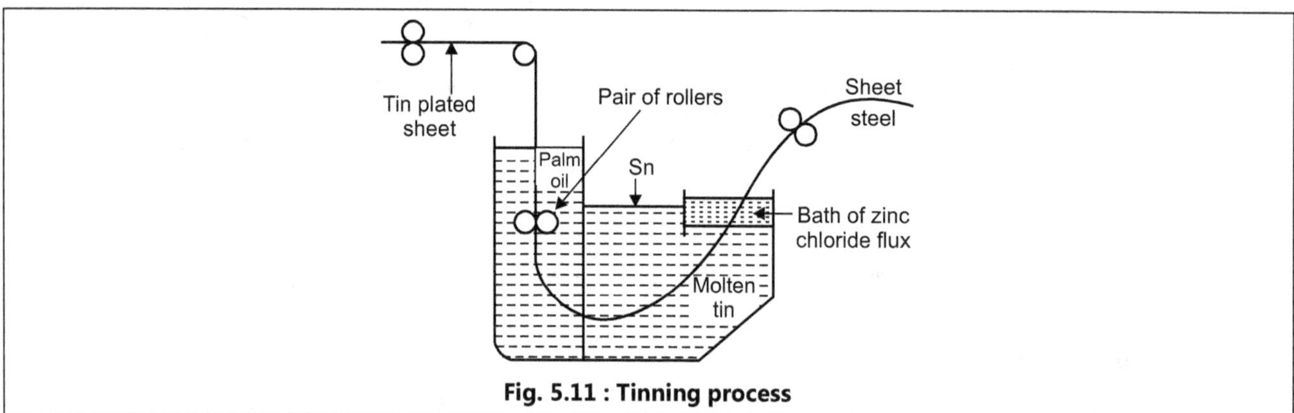

Fig. 5.11 : Tinning process

The process of tinning is carried out as follows :

- The sheet of steel to be tinned is first cleaned by dilute sulphuric acid to remove the oxide film and the impurities, if any.

- Then it is dipped in a bath containing molten flux of zinc chloride, which helps the molten metal to get adhered to the metal sheet.

- Then it is dipped in another bath containing molten tin. Finally, it is dipped in a suitable vegetable oil, to protect the hot tin coated surface against oxidation. It is then passed through a series of hot rollers. The rollers remove the superfluous and make the coating uniform all over the surface of the metal sheet.   (Fig. 5.11 and 5.12).

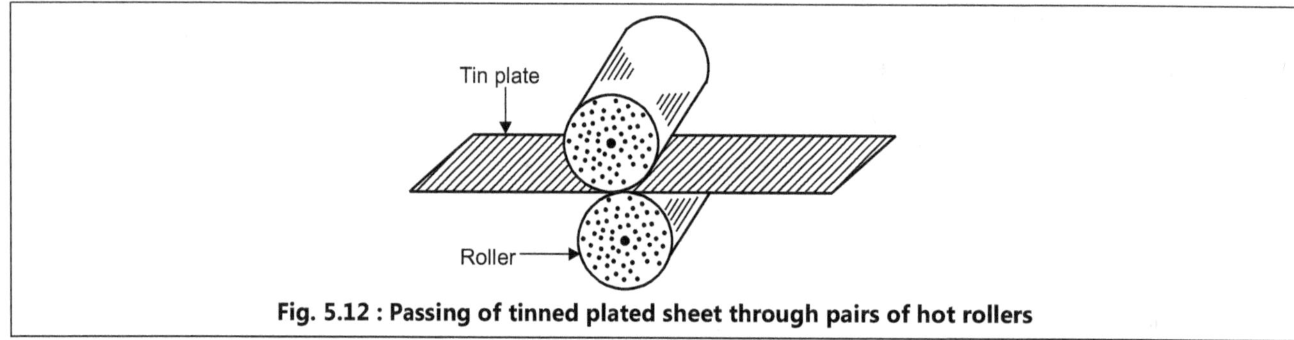

Fig. 5.12 : Passing of tinned plated sheet through pairs of hot rollers

Applications of Tinning Process

(i)   Tinning is used in manufacturing of various types of cans for storing food-stuffs, biscuit tins, kitchen utensils, oil, ghee, pickles, medicines, kerosene etc., because tin protects the metal from corrosion and prevents food poisoning.

Tin-plated steel sheets are used for making trunks, boxes, for roofing, for vessels for storing petroleum etc.

(ii) Copper wire to be insulated with rubber is first tinned to protect it from sulphur attack of the rubber.

(iii) Copper wire is also tinned to facilitate soldering.

(iv) Tinned copper sheets are used for cooking utensils and refrigeration equipment. Copper tubes are used in refrigeration.

Distinction between Galvanising and Tinning :

Galvanising	Tinning
1. It is the process of covering iron or steel sheet with a thin coat of zinc to prevent it from rusting.	1. It is the process of covering mild steel sheet with a thin coat of tin to prevent it from corrosion.
2. In galvanising, zinc protects the iron as it is 'more' electropositive than iron and does not allow iron to pass into the solution.	2. In tinning, tin protects the base metal from corrosion as it is 'less' electropositive than iron, therefore, it is more resistant to chemical attack.
3. In galvanising, zinc continues to protect the metal by galvanic cell action even if the coating is broken.	3. In tinning, tin protects the metal till the coating is perfect, a break in coating causes rapid corrosion.
4. Galvanised ware (zinc coated-utensils) cannot be used for storing food-stuffs which are acidic in nature, as zinc reacts with acids forming zinc compounds which are poisonous.	4. Tin coated utensils can be used for storing food-stuffs as tin protects the metal from corrosion and prevents food poisoning.

2. Metal spraying :

- It is the process of spraying of molten metal on the surface of base metal by spraying gun or pistol to provide a coating.
- Metal spraying is the most recent methods of developing protective coating.

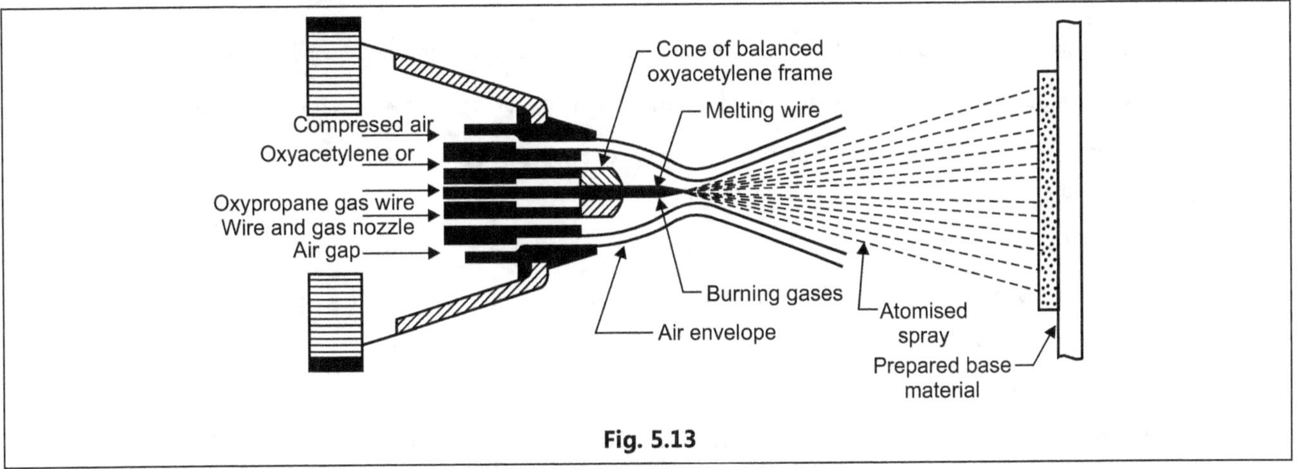

Fig. 5.13

Procedure :

- The spraying gun consists of a duct for compressed air and is fitted with oxyhydrogen flame.
- The coating metal in the form of wire is fed into a gun.
- The metal wire gets melted by oxyhydrogen flame and atomised using compressed air to form a fine spray.
- The spray can be directed to the surface to be coated (base metal), where fine molten droplets rapidly solidify and form the coating. This coating protects the base metal from the action of the surrounding medium.
- The process is not so effective as the coating may be porous and less adherent.
- This method is advantageous and handy from the point that the thickness of the coating can be controlled and irregular surfaces can be coated efficiently.

Applications of metal spraying process :

(i) Sprayed coating can be applied to non-metallic bases made of wood, plastic and glass.

(ii) Coating can be applied to fabricated structure and there will be no possibility of damage of the coating during the assembly of parts.

(iii) Worn out machine parts can be reclaimed by metal spraying.

(iv) In chemical industry, the coating of metals like Al, Zn, Ni, Sn, Pb, etc. is made by the method of spraying.

3. Electroplating :

(Nov. 18)

- Electroplating is the best and common method of producing metallic coating of more resistant metals (like silver, gold, nickel and chromium) on the base metal with the help of electric current.

- The purposes of electroplating are :

  (i) Decoration, (ii) Protection.

(i) Decoration :

- Coating of a superior metal over an inferior metal (base) is done in order to have beautiful and attractive appearance.

- It increases the commercial and decorative value of the article.

(ii) Protection :

- A coating of more resistant metal like Zn, Cr, Ni, Cu etc. is applied on a base metal like iron by electroplating in order to save the base metal from rusting and corrosion.

Process :

- The surface of the article to be electroplated is first thoroughly cleaned and suspended into the electrolyte and made as cathode.

- The anode consists of the pure metal whose coating is desired on the article. It is also suspended in electrolyte solution.

- The electrolyte generally consists of a salt solution of the coating metal together with certain other substances.

- On passing electric current, the metal ions from electrolyte get deposited on the article. The equivalent amount of anode gets dissolved in the form of ions and passed into the electrolyte.

- A thin layer of superior coating metal is obtained on the article (cathode).

- The smooth and brighter the deposits obtained depend upon the favourable conditions such as low temperature, high current density and high metal ion concentration.

Applications of electroplating :

- The method of electroplating is used for protective coating of metals and non-metals.

- Electroplating is used to improve appearance and hardness of metal.

- It is used to increase resistance to corrosion, wear and chemical attack.

- This method is used for coating of non-metals like wood, glass etc. to impart decoration, preservation and strength.

- It is also used in making surface conductive and utilization of light weight, as in case of wood and plastic.

4. Metal cladding :

- It is the process by which a dense, homogeneous layer of coating metal is bonded firmly and permanently to the base metal on one or both sides.

- In this process, the base metal to be protected against corrosion is sandwiched or cladded between two sheets of coating metal.

- This sandwich is then passed through two heavy rollers maintained at high temperature.

- The sandwiched (cladded) metal becomes cathodic with respect to base metal (anodic) so that the electrolytic protection is provided.

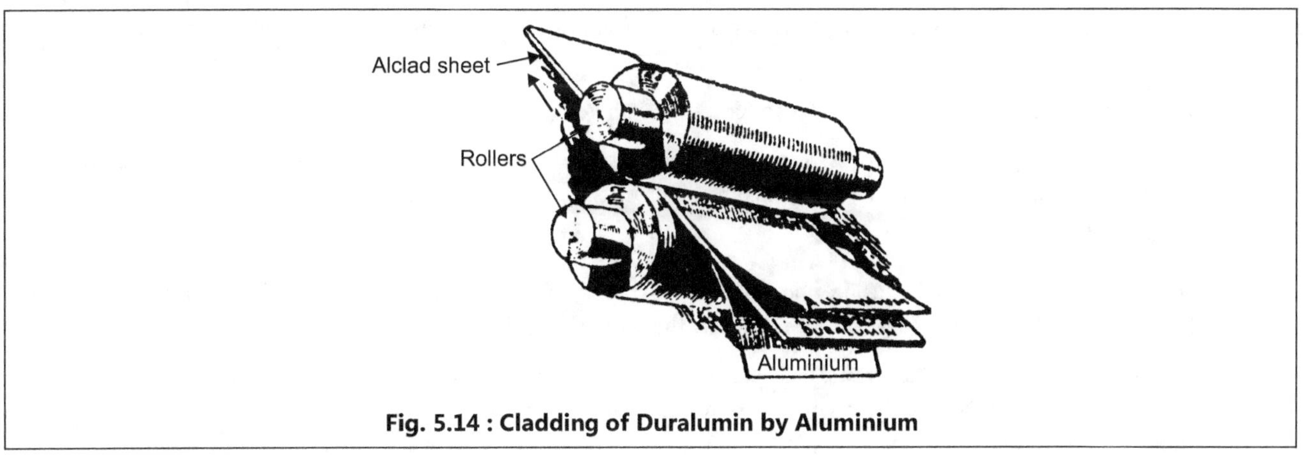

Fig. 5.14 : Cladding of Duralumin by Aluminium

- The choice of cladding material depends upon the resistance required for any particular environment. Nearly all existing corrosion resisting metals like nickel, copper, lead, silver, platinum and alloys like stainless steel, nickel alloys, lead alloys can be used as cladding materials.

- Among the base metal on which cladding is done are mild steel, aluminium, copper, nickel and their alloys.

Applications of metal cladding :

1. This method is widely used in aircraft industry where 'alclad' sheets are used. In alclad, 99.5% pure aluminium sheets are cladded by sandwiching the duralumin sheet. (Refer Fig. 5.13)

2. Aluminium cladding is also obtained by rolling the steel sheet at about 400°C. This results in a good bond between aluminium and steel. Subsequent annealing at 550°C causes aluminium to unite with iron producing $FeAl_3$.

3. Copper-clad steel wire is made by forcing a steel rod into a closely fitting copper tube. The wire is used widely for electrical conductors, processing, combining strength of steel with high conductivity of copper.

Advantages :

1. To develop surface properties like corrosion resistance in steel sheets.

2. To produce combining advantage of strength due to steel wire and electrical conductivity due to copper by producing copper-cladded steel wire.

3. By producing alclad sheets (i.e. aluminium is cladded on duralumin sheets), the light and strong alloy required in aircraft industry can be obtained.

4. The cladding metal (i.e. coating metal) provides the electrolytic protection for the base metal.

Disadvantages :

1. By metal cladding only plain surfaces can be protected.

2. Cladding is not perfect, the irregular surfaces provide galvanic cell action in the presence of moisture. Hence, the corrosion cannot be absolutely prevented by this method.

5. Cementation or Diffusion coating :

- The diffusion coating is obtained by heating the base metal in a revolving drum containing a powder of the coating metal.

- Diffusion of coating metal into base metal takes place, resulting in the formation of layers of alloy of varying composition.

- The layer just adjacent to the base metal may be an intermediate compound, a solid solution. The outer layers are richer in the coating metal.

- The coating thickness is controlled by varying the time of treatment and temperature.

- This process is suitable for coating small articles of uneven surfaces and intricate shapes like bolts, screw, threaded parts, valves etc.

- The coating metals used are those which can form alloy with iron (like Zn, Al, Cr etc.)

- When coating metal is Zn, the process is called as sherardizing, when coating metal is chromium, the process is called chromizing and when coating metal is aluminium, the process is called colorizing.

(a) Sherardizing :

- It is the process of cementation using zinc powder as a coating metal. This process is used especially for coating of odd shaped iron articles with thin layer of zinc to protect them from corrosion.

- The iron articles (such as bolts, screws, nails etc.) to be coated are first cleaned and then packed with zinc dust and zinc oxide (ZnO) powder in a steel drum which is provided with electrical heating arrangement to raise the temperature to about 400°C. It is rotated by means of a motor.

- The drum is slowly rotated for 2-3 hours, at 350-400°C.

- During this process, zinc gets diffused into iron forming Fe-Zn alloy at the surface which protects the iron surface from corrosion.

- The main advantage of sherardizing is that coating is quite uniform even if the surface has crevices or depressions and there is practically no change in the dimension of articles.

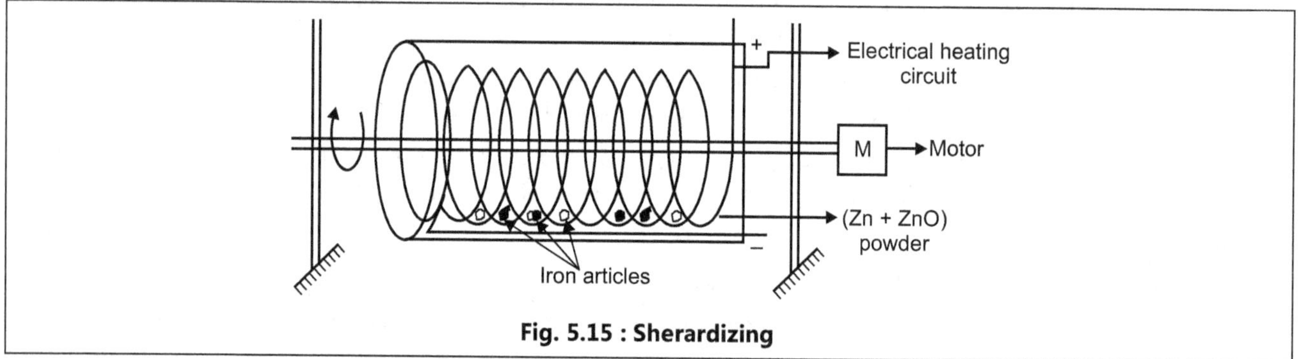

Fig. 5.15 : Sherardizing

Application :

- Sherardizing is used for protecting small steel articles like bolts, screws, nuts, threaded parts, washers, valves and gauge tools etc. against atmospheric corrosion.

(b) Colorizing :

- It is carried out by first sand-blasting the metal objects and then heating them in a tightly packed drum with a mixture of aluminium powder and Al_2O_3 together with a trace of NH_4Cl as a flux.

- The process is done in absence of air and sometimes in reducing atmosphere of hydrogen.

- The layer formed has an approximate composition of Al_3F_2 corresponding to about 25% Al by weight.

- Colorizing is applied for protection of furnace parts.

(c) Chromizing (coating of Cr) :

- It is carried out by heating together a mixture of 55% chromium powder and 45% alumina, together with base metal parts at about 1300 - 1400°C for 3 to 4 hours.

- The use of alumina prevents the coalescence of chromium particles.

- Chromizing is also produced by the interaction of a mixture of volatile $CrCl_2$ and H_2 with steel parts at about 1050°C.

- The diffusion of chromium into iron surface is more rapid than the above powder methods. The process occurs in three stages.

$$Fe + CrCl_2 \rightleftharpoons FeCl_2 + Cr \qquad \text{(Displacement)}$$

$$CrCl_2 + H_2 \rightleftharpoons Cr + 2HCl \qquad \text{(Reduction)}$$

$$CrCl_2 \rightleftharpoons Cr + Cl_2 \qquad \text{(Thermal decomposition)}$$

- The corrosion resistance of chromized coatings corresponds to the ferrite stainless steels.

- The chromium content in the diffusion method varies generally from 10 to 20%.

- The layers are supposed to be solid solutions of iron and chromium. Chromizing process is fairly extensively applied for the protection of gas turbine blades.

Distinction between Galvanizing and Sherardizing :

Galvanizing	Sherardizing
1. It is the process of coating iron or steel sheet with a thin coat of zinc by hot dipping.	1. It is the process of coating small iron or steel article by alloying at the surface of base metal (iron).
2. This process is carried out in large tanks by dipping iron sheet in a bath of molten zinc at a temperature of about 425-460°C by covering the bath with a flux of ammonium chloride.	2. This process is carried out in rotating closed drum like furnace by packing the small iron or steel article in zinc powder at a temperature of about 350-400°C in reducing atmosphere.
3. In galvanizing, a layer of Zn–Fe alloy is formed to which the outer layer of zinc sticks.	3. In sherardizing, zinc gets diffused into iron forming Fe-Zn alloy at the surface only.
4. This process is widely used for protecting iron exposed to the atmosphere, as in the case of roofs, wire fences, pipes and fabricated articles from galvanized sheets like buckets, tubes, etc.	4. This process is used for protecting small iron and steel articles like bolts, screws, nails, nuts etc. There is very little change in dimensions of small articles due to the formation of very thin layer of zinc.

PRACTICE QUESTIONS

1. Define corrosion. Mention the types of corrosion.

2. Explain the magnitude of corrosion problem.

3. Define oxidation corrosion. Write its mechanism.

4. Define wet or electrochemical corrosion. Explain its mechanism.

5. Explain the mechanism of oxidation when corrosion is brought about by the direct action of oxygen on metals in the absence of moisture.

6. Explain the role of oxide film formed during corrosion.

7. (i) Distinguish between the corrosion of magnesium and aluminium caused by the atmospheric oxygen.

   (ii) Why aluminium does not corrode fast ?

8. Explain hydrogen evolution mechanism of wet corrosion. When does it occur ?

9. Write short notes on :

   (i) Concentration cell corrosion.   (ii) Pitting corrosion.

   (iii) Waterline corrosion.   (iv) Crevice corrosion.

10. Explain differential aeration by absorption of oxygen mechanism.

11. Discuss the factors affecting rate of corrosion.

12. List the different methods used for the prevention of corrosion.

13. Wet corrosion is based on galvanic cell action. Explain in support of this statement.

14. "In any structure, two dissimilar metals under humid conditions should not be in contact with each other". Why ?

15. (i) Define galvanizing. State the principle of this method. Write its advantages and disadvantages.

    (ii) Describe the process of galvanizing for the protection of metal from corrosion.

16. Define tinning. Describe the process with neat labelled diagram. Write its applications.

17. Distinguish between galvanizing and tinning.

18. Write explanatory notes on the following :

    (a) Galvanising, (b) Sherardizing, (c) Tinning, (d) Electroplating, (e) Spraying, (f) Metal cladding, (g) Colorizing, (h) Chromizing.

19. Explain the difference in the following pairs of processes adopted to protect iron and steel against corrosion.

    (a) Galvanising and sherardizing.   (b) Tinning and tin plating.

    (c) Dipping and spraying.   (d) Galvanising and tinning.

20. (a) Which properties of zinc make it useful for galvanising ?

    (b) Which properties of tin make it useful for tinning ?

21. "Zinc coated steel is not useful but tin coated steel is useful for storing and canning food." Why ?

22. Discuss the factors affecting the rate of electrochemical corrosion.

23. List the different methods used for the prevention of corrosion.

24. Give different types of metallic coatings. Explain how the following processes are carried out :

    (i) Metal spraying.   (ii) Tinning.

25. Describe the process of sherardizing for the protection of metals from corrosion.

26. Define corrosion. Explain the role of metallic impurities in galvanic corrosion. Give its advantages.

27. What is metal cladding ? Describe the process for the protection of metals from corrosion and give its disadvantages.

28. (a) Name and describe the method used for coating large and irregularly shaped articles, for prevention of corrosion.

    (b) Explain the process of spraying. Write four advantages of this process.

29. (a) Describe the process of tinning for the protection of metal from corrosion and give its applications.

    (b) Define tinning. State and explain the principle of tinning process.

PAINTS AND VARNISHES

(A) PAINTS

6.1 DEFINITION   (Nov. 18)

- *"Paint is a mechanical dispersion mixture of one or more pigments in a vehicle".* The 'vehicle' is a liquid consisting of a non-volatile film forming material, 'drying oil' and a highly volatile solvent, 'thinner'. When a paint is applied to a metal surface (usually by brushing or spraying) the thinner evaporates, while the drying oil slowly oxidises forming a dry pigmented film. To accelerate the drying process of oil, small amount of catalysts (called driers) are also added to the paint.

6.2 PURPOSES OF APPLYING A PAINT

- Paints are generally used by builders / engineers for covering the surfaces of buildings and engineering materials such as wood, iron etc. The main purposes of applying paint are as follows :

    1. Paint protects iron and wood from wear and tear.

    2. Paint protects wood from different types of insects, fungus and moisture.

    3. Paint protects iron from rusting and corrosion.

    4. The painted surface reflects heat and light nicely.

    5. The painted surface provides a smooth and beautiful appearance.

6.3 CHARACTERISTICS OF A GOOD PAINT

- A good paint should possess the following characteristics :

    1. It should be fluid enough to be spread easily over the surface to be protected.

    2. It should possess high covering power.

    3. It should have brushing characteristics. Brushes may be made according to the type of paint to be applied.

    4. It should form a quite tough, uniform, adherent and durable film.

    5. It should protect the painted surface from corrosion effects of the environment.

    6. Its film should not get cracked on drying.

    7. Its film should be washable.

    8. Its film should be glossy.

6.4 PRINCIPAL CONSTITUENTS AND FUNCTIONS OF A PAINT

- The important constituents of a paint are :

    (1) Pigments, (2) Vehicle or medium (drying oils), (3) Thinners, (4) Driers, (5) Fillers or extenders and (6) Plasticizers.

    1. Pigments : *"Pigment is a solid substance, which forms a paint when mixed with drying oil".* It is the principal constituent forming the body of a paint.

Types of pigments :

   (a) White pigments : White lead [$2PbCO_3.Pb(OH)_2$], zinc oxide (ZnO), lithophone ($BaSO_4 + ZnS$), titanium oxide (TiO_2) etc.

   (b) Coloured pigments :

   (i) Red - Red lead, ferric oxide, chrome red, venetian red.

   (ii) Green - Chromium oxide (Cr_2O_3).

(iii) Blue - Prussian blue $Fe_4[Fe(CN)_6]_3$, ultramarine blue (sodium aluminium silicate + sulphide), cobalt blue (70% Al_2O_3 + 30% Co_2O_4).

(iv) Black - Carbon black, lamp black.

(v) Brown - Brown amber.

Characteristics of good pigments : Good pigments should be (i) Cheap, (ii) Opaque, (iii) Chemically inert, (iv) Non-toxic, (v) Freely mixable with film-forming constituents of the oil.

Functions of pigments : The functions of pigments in paint are to :

1. Provide opacity and colour to the paint film.

2. Give an aesthetical appeal (i.e., pleasing to look at) to the paint film.

3. Give strength to the film.

4. Give protection to the paint film by reflecting harmful ultraviolet light.

5. Provide resistance to paint film against abrasion or wear.

6. Improve the impermeability of paint film to moisture.

7. Increase weather resistance of the film.

2.   Vehicle (or medium or drying oil) : *"These are the film-forming constituents of paint."* In oil paints, the vehicle, or medium includes drying oils such as linseed oil, tung oil, dehydrated castor oil and semi-drying oils such as soyabean oil, fish oil, rosin oil etc. The semi-drying oils are slow-drying and hence are used as admixture with drying oils.

Functions of vehicle or medium : Drying oils supply paint film with :

1. Mainly film forming constituents

2. Toughness

3. Adhesion

4. Durability and water proofness

5. Protective film by evaporation or by oxidation and polymerisation of the unsaturated constituents of the oil.

3.   Thinners : "*These are the volatile liquid substances which evaporate easily after application of paint.*" Thinners are added to paints to reduce the viscosity (consistency) of the paint, so that they can be easily applied to the metal surface. Common thinners used in paint are :

(i)   Aliphatic hydrocarbons : Mineral spirits, naphthas and other petroleum fractions (e.g. kerosene).

(ii) Aromatic hydrocarbons : Benzene, tuluol, xylol, methylated naphthalene.

(iii) Turpentine and dipentene.

Functions of thinner : The important functions of thinners are as follows :

1. They suspend pigments.

2. They dissolve film-forming materials.

3. They thin (or to reduce viscosity of) the concentrated paints for proper handling and to impart better covering power.

4. They help the drying of film by evaporation.

4.   Driers : *"These are oxygen carrier catalysts."* They accelerate the drying of oil film through oxidation, polymerisation and condensation. The commonly used driers are heavy metallic soaps such as napthenates, linoleates, resinates and tungstates of heavy metals Co, Mn, Pb and Zn.

(i) Cobalt substances are the most efficient of all and are 'surface-driers' (ii) Lead substances are 'bottom driers', while (iii) Manganese substances are 'thorough driers.' Too much of a drier tends to produce hard and brittle paint films.

Functions of Driers :

1. They improve the drying quality of oil-film.

2. They act as oxygen carrier catalysts.

3. They accelerate the drying of oil film by oxidation, polymerisation and condensation.

5.   Fillers or Extenders : These are the inert materials which improve the properties of the paint although they have low opacity. Commonly used extenders are barytes ($BaSO_4$), $CaCO_3$, $CaSO_4$, talc, asbestos, gypsum, clay, chalk, silica, magnesium silicate ($MgSiO_3$), slate powder etc.

Functions of extenders :

1.   They reduce the cost of a paint.

2.   They increase the durability of a paint.

3.   They help to reduce the cracking of dry paint and keep the pigments in suspension.

4.   They serve to fill the voids in the film.

5.   They increase the random arrangement of pigment particles.

6.   They act as carriers for the pigment colour.

6.   Plasticizers : Sometimes, plasticizers are used in paints (i) to give elasticity to the film, (ii) to prevent cracking of the film. Plasticizers in common use are triphenyl phosphate, tricresyl phosphate, diamyl phthalate, tributyl phthalate, and dibutyl tartarate.

6.5 METHODS OF APPLICATION OF PAINTS   (Nov. 18)

•   Paint is applied to the surface by many methods. The principle methods used are the following :

1.   Brushing : An easy method of applying paint is by dipping a brush in the paint and then applying on the surface. A good brush should have fine bristles so that its marks left after painting, diffuse in each other and gives a smooth finish on the painted surface. A highly thick (or viscous) paint is difficult to be applied by brushing. Hence, such paints are mixed with proper amount of thinner (e.g. turpentine) to get required consistency for brushing.

2.   Spraying : This method involves the atomising of paint with the help of a spray-gun. This method is quite quick and also gives a smooth painted surface, a lot of labour is also saved. But a lot of paint is wasted and the surrounding atmosphere gets polluted by fine suspended particles of paint in its air. Recently, these disadvantages have been eliminated by making the article to be sprayed as positive, whereas the spray-gun is given a negative charge. The paint spray, as it comes out of the spray-gun, are negatively charged and immediately attracted towards the article to be painted. Thus, the chances of pollution are eliminated.

3.   Dipping : This method is used for painting the objects having uniform surfaces (i.e., free from depressions, pockets and crevices etc.). In this method, the article is dipped in a tank containing paint. It is then taken out, drained (to remove excess of paint) and finally dried. This method has two disadvantages :

(i)   The thickness of the painted film is not uniform.

(ii)   The tears of paint film at some points and formation of paint beads are caused (due to dripping).

However, these defects can be removed by passing painted articles over a grid, carrying a high-voltage charge. Due to this, the excess of paint gets repelled, thereby giving a uniform smooth surface.

4.   Tumbling : It is applicable for painting small-sized wooden articles. They are put in a rotating barrel, containing a proper quantity of the paint. The barrel is then closed and rotated for half an hour with the help of a motor. Paint gets deposited over the articles, which are then taken out and dried.

5.   Roller coating : It is applied when the article is in the form of flat sheets. These are passed through rubber-coated rollers, while a little paint falls over the rollers. During their passage through rollers, the articles are painted uniformly.

6.6 FAILURE OF A PAINT FILM

A paint may fail to wear out due to several causes which are given below :

1.   **Chalking :** "It is the progressive powdering of the paint film on the painted surface." This occurs due to improper dispersion of pigments, vehicle or by destruction of binder by the continuous exposure to light.

2.   **Flaking (or Peeling) :** "It is the peeling of the paint film from the painted surface." This is due to the presence of dust particles or greasy matter in the paint. These foreign matters result in poor adhesion of the paint to the surface, thereby causing the peeling off.

3. **Cracking :** The cracking of paint film is due to unequal expansion and contraction of different coats of the paint on variation of the temperature of the exposed film. This defect, however can be prevented by making the primary (first) coat a harder one.

4. **Blistering :** It is due to the presence of moisture in the article, particularly wooden surfaces. The inherent dampness, after some time, comes to the top of surface below the paint film, thereby causing blisters. This defect can, however, be prevented by applying the primary coat of some dehydrating agent.

5. **Change of colour of paint film (or bleeding) :** When paint film is exposed for a longer period, the chemical effect of atmospheric gases takes place on the paint. e.g., white lead paint film tarnishes (i.e., turns blackish) when exposed to the atmosphere containing H_2S. Similarly, ZnO paint film turns yellow.

Prevention of failure of paint film : The failure of paint film can be prevented by :

1. Proper mixing and proportioning of ingredients of proper characteristics.

2. Only using a suitable paint for a particular job.

3. Carefully preparing the surface before application of paint.

4. Applying a suitable primary coat.

5. Evenly applying the paint by a suitable method (e.g. brushing, spraying etc.)

6. Sufficiently, allowing each paint coat to dry, before the next coat is applied.

(B) VARNISHES

6.7 CHEMICAL RESISTANT COATINGS

- Our chemical resistant paint and coatings offer protection to concrete, cementitious mortars and rendering, metal, timber, asbestos and more in a wide range of chemical processing environments and industries. Protection against chemicals comes in various guises around the workplace every day, from protective clothing, storage of chemicals, protecting bundled areas from accidental spillage, etc. These products help provide a solution for our customers to protect different surfaces where necessary from a vast range of chemicals, whether it be chemical and acid resistance for floor areas, tanks (internally and externally), bunds, walls, and more. The superior technology used in our products means that even some of the harsdest chemicals at high concentration can be protected against with minimal effort and cost.

- For chemically resistant tank and pipe linings. rawlins paints have set up a specialist category, which focuses on a range of metal and concrete coatings for the marine, offshore and processing plant sectors.

Products :

- Nuclear facilities.

- Traffic and industrial fumes.

- Liquid spillages.

- Excellent flexibility.

- Improved resistance against chemical attacks to a surface or substrate from long-periods of storage and transportation - typically from two-component coatings, like epoxy, polyurethane, methacrylate, and vinyl ester resins.

- Corrosive petrol, diesel, acids, alkalis and solvents in industrial processing facilities where minimal surface cleaning is attainable.

- Enhanced surface lubrication is required.

- Improved component life from durable topcoats.

- Surface heat and low-temperature handling capabilities.

- Storage of high temperature chemicals.

- Extreme protection in high humidity areas where high rates of salt water corrosion may be typical.

- Improved resistance to anti-galling lubricants.
- Protects marine environment surfaces from :
  - Salt spray.
  - Excessive moisture.
  - Oxidation and rust.
  - Other environmental chemicals.
- Resilience against chemical compounds and corrosive materials.
- Improved colour retention.
- Extended surface and substrate lifespan.
- High-gloss finishes.
- Cost-effective long-term coating solutions for flooring in heavy machinery plants.
- Abrasion resistance.
- Thin film coating requirements that offer extended protection to the underlying surface or substrate.
- And more.

The chemical resistance of a coating will depend on a variety of criteria, including:

- The type of chemical.
- Its concentration.
- Temperature - environment and substance.
- Level of exposure - full/partial, occasional contact.
- Length of time exposed to the chemical.
- The full paint system specifications.
- Has there been any maintenance or repair work conducted?
- Interior or exterior location.
- Substrate type :
  - Concrete.
  - Metal.
- Submerged, underground / under-soil location.
- Will the surface be subject to abrasion and vehicular use?
- Cleaning regimes.
- Humidity.
- Required lifespan.
- Substrate or surface preparation.
- Stain resistance.
- Weather protection characteristics.
- How porous is the substrate?
  - Surface exposure rate.
  - Walls.
  - Floors.
  - Ceilings.

6.8 HEAT RESISTANT PAINTS

- Heat resistant paints available at Rawlins Paints are designed to withstand high-temperatures of upto 750°C on commercial and industrial metal/metallic surfaces and structures - ranging from heat treatment applications, including incinerators and furnace chimneys, to pipework and vessels in chemical processing plants. For very demanding inland/offshore environments, VHT (very-high-temperature) paints and coatings are also required to deliver high-performance humidity resistance and anti-corrosive protection against liquids including water, oil and gasoline/petrol, as well as salt, gas, rust and chemical solvents.

- Heat resistant paints and coatings are not designed to deliver fire retardant properties but can be used on industrial chimneys and steel melting applications. These high-temperature coatings will not react to flames and cannot contain the surface spread of flames. The most commonly used heat resistant paint colours include black and silver (also referred to as aluminium). Some heat resistant coatings we stock are also available in white, grey, red oxide and orange.

- High-temperature paint is designed to not break down at very high temperatures, so they can be safely used for industrial applications and processing equipment that need to be regularly elevated at temperatures of upto 750°C.

- Fire retardant paints, on the other hand give off an inert gas to slow down and stop the spread of fire on the surface and intumescent paints provide an insulating barrier by swelling upto 40 times their original thickness.

  Surfaces, industrial environments and applications for which heat resistant paints are suitable for, include:

  - Incinerators
  - Kilns
  - Flare stacks
  - Pipework
  - Vent covers
  - Petrochemical industry structures and refineries
  - Chemical plants
  - Natural gas processing and biochemical plants
  - Power stations
  - Offshore gas and oil rigs
  - Metal, aluminium alloy, cast iron, etc. dividing and collecting breeching used on:
    - Water pipes and hoses
    - Dry riser inlets
    - Fire hose coupling
    - Fire hydrants
    - Stand pipes
    - Hydrant stand posts
    - Landing valves
  - Process vessels
  - Engines
  - Furnace chimneys
  - Chimneas
  - Radiators
  - Boiler fronts
  - Grills, exhaust systems and exhaust pipes

- High-temperature steelwork
- BBQs
- Grates
- Steam pipes
- Metal fireplaces
- Engine headers
- Smokestacks
- Ductwork
- Valves
- Sampling lines

How to Apply Heat Resistant Paint

One of the more popular ways to apply heat proof paint is via spray devices. Jotun's range can be applied with spray guns also available at Rawlins Paint.

Very High Heat Resistant Metal Paint

Various metal structures, surfaces and substrates that get up to high-temperatures require painting, and finding the best heat resistant paint for your needs is key to ensuring the finish is appropriate for use.

If you are looking for high heat paint for barbecues, boilers, exhaust pipes and other smaller interior surfaces, then have a look at our Rust-Oleum and Blackfriar heat resistant paint.

Rust-Oleum Heat Resistant Paint 750°C can be used on bare metal or slightly rusted surfaces and provides a heat resistant coating. Rust-Oleum HR protects upto a dry heat of 750°C and can be applied inside or in a sheltered outdoor environment. Perfect for engines, barbecues, vents and pipes – it is available as an aluminium heat resistant coating or in black satin-matt.

Enamel Paint

An enamel heat resistant coating, Coo-Var Heat Resistant Satin Black can be used on surfaces that need to withstand heat upto 600°C and is ideal for areas that frequently get hot. It can be used as a single coat for hot pipes, radiators and fire surrounds.

Another high-quality heat resistant enamel paint is Teamac High Temperature Aluminium Paint. Teamac can be used at temperatures upto 450°C and is applied with a brush or roller. The satin finish is ideal for metal and can also be used as a single coat on fire surrounds and hot pipes.

Aluminium Paint

With a heat resistance of upto 250°C, Jotun Aluminium Paint HR is perfect for engine parts and exhaust pipes. Its aluminium finish coating can be used as a heat resistant primer, mid coat or top coat and can be used on carbon steel and aluminium substrates if they are properly prepared.

Industrial Heat Retardant Paint

Many heat resistant coatings also have an anti-corrosive finish, which adds more protection for the surface. Rust-Oleum 4200 Heat Resistant Primer can be used on incinerators, stacks, kilns and other very high temperature surfaces that are prone to rust. This heat proof primer should be used on new, bare steel or blasted steel underneath Rust-Oleum 4200 Heat Resistant Topcoat.

If you are looking for protection from corrosion on stainless steel industrial surfaces, Jotun Jotatemp 650 is ideal. Jotatemp 650 works specifically on surfaces that need to withstand extreme high and low temperatures, between –185°C and 540°C and can be applied directly onto a surface that is at 400°C.

For large aluminium surfaces, such as offshore structures, petrochemical and chemical plants and power stations, International Intertherm 891 is an excellent choice. Intertherm 891 is a general-purpose heat resistant paint that gives a silver finish as it is coloured with aluminium flake. It can be used as an industrial maintenance coating on steelwork with a temperature ranging from ambient to 315°C.

6.9 CELLULOSE SPRAY PAINTS

- In order to help people understand the specific paint types that we stock, we have created a series of dedicated blogs exploring each of these paints. These blogs aim to provide information about the specific characteristics of these paints, their benefits, and also the appropriate safety procedures that should be observed when working with each of these paint types.

- In this blog, we will look specifically at cellulose paints. Historically, this paint type was used utilised widely in the automobile industry. This is perhaps unsurprising, since cars require incredibly hard paintwork in order to survive the general wear and tear of their usage and cellulose paint coatings, once cured provide an incredibly durable coating.

- However, due to the fact that cellulose paints are highly toxic, flammable and very fast drying, working with them can be extremely unforgiving. Especially if you are inexperienced with working with this paint type and spray paints in general. One way in which these shortcomings can be avoided is by making use of a cellulose based spray primer. This is because these spray based primers apply only a light dusting onto the surface that they are applied and dry within seconds. This is especially beneficial if you wish to use your cellulose paints with plastic surfaces, as in larger quantities with longer exposure, this paint can actually cause plastic to melt. However when applied as outlined above the paint dries before any damage can be done to the underlying plastic, enabling the advantages of Cellulose paints to be enjoyed on this material.

What are the ingredients of Cellulose Paints?

Pigment : This ingredient is what gives the cellulose paint its colour, it also successfully hides everything that is beneath the coating, providing opacity. In metallic paints, perhaps unsurprisingly, the pigment is provided through the inclusion of little pieces of metal within the paint itself. Equally, cellulose paints that are sold as primers, typically do not have a pigment included in them, therefore the paint is colourless.

A Binder or Vehicle: The binder is just as important as the pigment. This is the ingredient that ensures that the pigment particles actually stick to one another, while at the same time making sure that the paint successfully bonds to the surface to which it is applied. The strength and durability of cellulose paints can be attributed to the binder that is used to create them. This is because the binder is the factor that determine the overall hardness or flexibility of the coating once it has dried. Additionally, the binder is responsible for the level of chemical resistance provided by the coating. Often it is the binder that differentiates paint types from each other, for example, cellulose paints utilise a different binder to enamel paints.

The Liquid, Solvent, Carrier or Dispersent : This third ingredient of cellulose paint is sometimes called any one of the above four names, which can lead to confusion. The liquid or solvent can serve a number of purposes. In many cases it is utilised in order to thin the paint and to make it more liquid in consistency. Many paints, before the addition of this ingredient have a paste like consistency, making them unsuitable for easy application. Once added, the liquid makes it easier for the paint to be applied and also affects the drying time. The amount of liquid or solvent mixed to a paint determines its consistency. To demonstrate the greater the level of liquid, the thinner the coating of paint. Typically, cheaper paints have far higher levels of liquid added to them as this is by far the cheapest ingredient. In the case of cellulose paints, it is the cellulose that is added as this ingredient that gives the paint its name and qualities.

Additives : In addition to the three ingredients outlined above, cellulose paint may have a number of additional additives that are added to it. In many cases these are added in order to enhance the characteristics of the paint. These characteristics can include:

- To disperse the pigment of paint throughout the solution and to prevent it from settling at the bottom of the spray can.

- To change the surface texture of the paint once applied, to give it a finish that ranges from gloss to matt.

- Preservatives to lengthen the longevity of the paint prior to its usage and application.

- To adjust the surface tension of the paint and to improve the flow of application.

- To either increase or decrease the overall time that it takes the coating to dry and cure once it has been applied.

6.10 LUMINOUS PAINTS

Luminous paint or **luminescent paint** is a paint that exhibits luminescence. In other words, it gives off visible light through fluorescence, phosphorescence, or radioluminescence.

6.10.1 Fluorescent Paint

- Fluorescent paints offer a wide range of pigments and chroma which also 'glow' when exposed to the long-wave "ultraviolet" frequencies (UV). These UV frequencies are found in sunlight and some artificial lights, but they - and their glowing-paint applications - are popularly known as black light and 'black-light' effects' respectively.

- In fluorescence the visible light component - sometimes known as "white light" - tends to be reflected and perceived normally, as colour; while the UV component of light is modified, 'stepped down' energetically into longer wavelengths, producing additional visible light frequencies, which are then emitted alongside the reflected white light. Human eyes perceive these changes as the unusual glow of fluorescence.

- The fluorescent type of luminescence is significantly different from the natural bioluminescence of bacteria, insects and fish such as the case of the firefly, etc. Bio-luminescence involves no reflection at all, but living generation of light (via the chemistry of Luciferin).

- There are both visible and invisible fluorescent paints. The visible appear under white light to be any bright colour, turning peculiarly brilliant under black lights. Invisible fluorescent paints appear transparent or pale under daytime lighting, but will glow under UV light in a limited range of colours. Since these can seem to 'disappear', they can be used to create a variety of clever effects.

- Both types of fluorescent painting benefit when used within a contrasting ambiance of clean, matte-black backgrounds and borders. Such a "black out" effect will minimize other awareness, so cultivating the peculiar luminescence of UV fluorescence. Both types of paints have extensive applications where artistic lighting effects are desired, particularly in "black box" entertainments and environments such as theaters, bars, shrines, etc. Out-of-doors, however, UV wavelengths are rapidly scattered in space (waves are known to bounce off surfaces, outdoors the bounce is imperceptible) or absorbed by complex natural surfaces, dulling the effect (as a human you only see reflected waves of photons you call light, black matte is ultimate absorption at infinite angles thus enhancing any escaped wave of light ; thus dim becomes bright). Furthermore, the complex pigments will degrade quickly in sunlight.

6.10.2 Phosphorescent Paint

- Phosphorescent paint is commonly called "glow-in-the-dark" paint. It is made from phosphors such as silver-activated zinc sulfide or doped strontium aluminate, and typically glows a pale green to greenish-blue colour. The mechanism for producing light is similar to that of fluorescent paint, but the emission of visible light persists long after it has been exposed to light. Phosphorescent paints have a sustained glow which lasts for upto 12 hours after exposure to light, fading over time.

- This type of paint has been used to mark escape paths in aircraft and for decorative use such as "stars" applied to walls and ceilings. It is an alternative to radioluminescent paint. Phosphorescent paint is typically used as body paint, on childrens' walls and outdoors.

- When applied as a paint or a more sophisticated coating (e.g. a thermal barrier coating), phosphorescence can be used for temperature detection or degradation measurements known as phosphor thermometry.

6.10.3 Radioluminescent Paint

- Radioluminescent paint is a self-luminous paint that consists of a small amount of a radioactive isotope (radionuclide) mixed with a radioluminescent phosphor chemical. The radioisotope continually decays, emitting radiation particles which strike molecules of the phosphor, exciting them to emit visible light. The isotopes selected are typically strong emitters of beta radiation, preferred since this radiation will not penetrate an enclosure. Radioluminescent paints will glow without exposure to light until the radioactive isotope has decayed (or the phosphor degrades), which may be many years.

- Because of safety concerns and tighter regulation, consumer products such as clocks and watches now increasingly use phosphorescent rather than radioluminescent substances. Radioluminescent paint may still be preferred in specialist applications, such as diving watches.

(1) Radium :

Radioluminescent paint was invented in 1908 and originally incorporated radium-226. Radium paint was widely used for 40 years on the faces of watches, compasses, and aircraft instruments, so they could be read in the dark. Radium is a radiological hazard, emitting gamma rays that can penetrate a glass watch dial and into human tissue.

(2) Promethium :

In the second half of the 20th, century, radium was progressively replaced with promethium-147. Promethium is only a relatively low-energy beta-emitter, which, unlike alpha emitters, does not degrade the phosphor lattice and the luminosity of the material does not degrade so fast. Promethium-based paints are significantly safer than radium; the half-life of ^{147}Pm however, is only 2.62 years, it is therefore not too suitable for long-life applications.

(3) Tritium :

The latest generation of the radioluminescent materials is based on tritium, a radioactive isotope of hydrogen with half-life of 12.32 years that emits very low-energy beta radiation. The devices are similar to a fluorescent tube in construction, as they consist of a hermetically sealed (usually borosilicate-glass) tube, coated inside with a phosphor, and filled with tritium. They are known under many names - e.g. gaseous tritium light source (GTLS), traser, betalight.

Tritium light sources are most often seen as "permanent" illumination for the hands of wristwatches intended for diving, nighttime, or tactical use. They are additionally used in glowing novelty keychains, in self-illuminated exit signs, and formerly in fishing lures. They are favored by the military for applications where a power source may not be available, such as for instrument dials in aircraft, compasses, lights for map reading, and sights for weapons.

Tritium lights are also found in some old rotary dial telephones, though due to their age they no longer produce a useful amount of light.

6.11 EMULSION PAINTS

* Emulsion paints are those paints in which water is used in place of organic solvents as thinner. Emulsion paint is a emulsion of two phases, one of which is water and this paint can readily be diluted or surface forming material (a synthetic resin latex) is dispersed in water by dispersing agent acting as a binder. It may also have in it another vehicle or film forming constituent such as oil. The binder may thus be oil, oleoresinous varnish or other emulsifying binder. Pigments and extenders are dispersed in such an emulsion. In addition to these constituents, an emulsion paint may also have emulsifying agents, stabilizers, driers, antifoaming agents, preservatives etc.

6.11.1 Constituents of Emulsion Paints

1. **Pigments :** Commonly used are water dispersing pigments like titanium dioxide.
2, **Extenders :** Commonly used are mica, diatomaceous clay, silica and magnesium silicate.
3. **Rubber like resins :** Semi-solid polystyrene, polyvinyl acetate, styrene butadiene copolymer are used.
4. **Oleoresinous materials :** Oils used are readily dispersible such as linseed oil, while resins used are alkyl resins, ester gum polystyrene etc.
5. **Emulsifying agents :** These are used for the dispersion of (a) rubber like materials are complex phosphates like tetra-sodium phosphate, (b) organic type pigments are dioctyl sodium succinate, sodium lauryl sulphate.
6. **Stabilizers :** Proteins like dextrin, starch, water soluble gums, bentonite, casein, soya-protein etc. A stabilizer imparts chemical resistance to the emulsion. Moreover, the protein provides body, thereby improving brushing action.
7. **Preservatives :** These are added to prevent the decomposition of any protein and to eliminate the growth of fungus. Important preservatives are mercuric chloride, thymol and chlorothymol.
8. **Antifoaming agents :** These are added to check any excessive foam formation by the agitation of emulsion paints during their manufacture. Important antifoaming agents are kerosene and pine oils.
9. **Driers :** These are added to an emulsion paint, which contains oxidisable oils. They are Co, Mn, Zn resinates and phthalates etc.
10. **Volatile material** is mainly water. Aqueous ammonia is also used in some cases.

6.11.2 Advantages and Properties of Emulsion Paints

1. The paint can be readily thinned or diluted with water.
2. It is very easily applied with brush or roller.
3. These paints are quick drying.
4. Even an inexperienced painter can paint with such a paint.
5. They are free from fire hazard and have no objectionable odour.
6. It can be recoated many times and can be easily applied over existing coatings.
7. The surface over which such a paint is applied can be easily washed with water.

6.12 PROPERTIES AND APPLICATIONS OF SPECIAL PAINTS

(a) Exterior house paints : These paints generally have constituents such as pigments (ZnO, TiO_2, white lead etc.), extenders (talc, barytes, clay etc.), vehicle (e.g. boiled linseed oil) and thinners (e.g. mineral, spirits, napthas etc.). Coloured pigments are also added in varying amounts for a light tint.

(b) Interior wall paints : These paints are prepared by mixing pigments (e.g. white and coloured pigments), vehicle (e.g. varnish or boiled linseed oil) and resins (e.g. emulsified phenol formaldehyde resins and casein).

(c) Chemical resistant paints : These paints consist mainly of baked oleoresinous varnishes, chlorinated rubber compositions, bituminous varnishes and phenolic dispersions as chemical resistant materials in paint formulations.

(d) Luminous paints : These paints consist of phosphorescent paint compositions such as pigments (sulphides of Ca, Cd and Zn dispersed in spirit varnish), vehicle (chlorinated rubber, styrol etc.) and sensitiser for activation in U.V. region.

(e) Metal paints : These are coating applied on the metal surfaces or bodies for protection and decoration. The coating may be of barrier type or galvanic type. In barrier type, a protective barrier is formed between the surface coated and its surroundings. These consist of a pigment, vehicle, anticorrosive agents (e.g. zinc or chrome yellow), resins (e.g. alkyds, epoxy, polyamides, chlorinated rubbers and silicones etc.).

Alkyds resist weathering of metals and polyamides form tough film resistant to chemicals, chlorinated rubbers resist action of soaps, detergents and strong chemicals and silicons are added as heat resistant and water repellents.

(f) Cement paints : These paints are prepared by mixing white cement with colouring matter or pigments, hydrated lime and fine sand as inert filler. Cement paints are available in the form of powder of particular colour. The dispersion medium may be water or an oil depending upon the purpose of coating. In case of stone or brick structure, the dispersion medium is of water, but boiled linseed oil is used as the dispersion medium if the purpose of coating corrugated metal surface. Before applying cement paint, a primer coat consisting of a dilute solution of sodium silicate and zinc sulphate is necessary. Cement paints have marked water proofing capacity, give a stable and decorative film and do not require fresh application even in 4 to 5 years, if coated even on rough surface.

(g) Water paints or Distempers : Distempers are nothing but water paints. The ingredients of a distemper are : (i) Whitting or chalk powder (the base), (ii) glue or casein (the binder), (iii) colouring pigment and (iv) water (the solvent or thinner).

Advantages : Distempers : (1) are cheaper than paints and varnishes, (2) can be applied easily on plasters, cement concrete or wall surfaces in the interior of buildings, (3) durable, (4) give smooth and pleasing finish to walls, (5) have good covering power.

Disadvantages of distemper as compared to paints :

(1) It is less durable than paints.

(2) The coatings of distemper are usually thicker and more brittle than water paints.

(3) The distemper film can shrink and crack, if not applied uniformly.

(4) The film of distemper is porous and allows penetration of moisture through it.

Uses : Distempers are used as a finishing coat on :

(1) White wash surfaces of interior walls.

(2) Plastered surfaces of interior walls.

(3) External surfaces of brick-works, concrete etc. after adding a weather resisting compound at the time of mixing.

6.13 VARNISHES – DEFINITION   (Nov. 18)

- *"Varnish is a homogeneous colloidal dispersion, solution of natural or synthetic resins in oils or thinners or both."*
- It is used as a protective and decorative coating of suitable surfaces and it dries up by evaporation, oxidation and polymerisation of its constituents, leaving behind a hard, transparent, glossy, lustrous and durable film.

6.14 CHARACTERISTICS OF A GOOD VARNISH

- A good varnish should possess the following characteristics :

1. It should be soft and tender.

2. It should dry very quickly.

3. On drying, the film of the varnish should be glossy, shining and aesthetically appealing.

4. It should give a hard, tough, durable wear and tear resistant film.

5. It should be able to adopt itself to expansion or contraction of coated material such as wood, because of variation of temperature.

6. Its film should be elastic, which does not crack or peel off or shrink on drying.

7. Its film should not fade or change colour on exposure to atmospheric weather.

6.15 CONSTITUENTS OF VARNISHES

Constituents of Classic Varnish :

(1) Drying oil :

There are many types of drying oils, including linseed oil, tung oil and walnut oil. These contain high levels of polyunsaturated fatty acids.

(2) Resin :

Resins that are used in varnishes include ambere, dammer, copal, rosin (pine resin), sandarac, balsam and others.

(3) Turpentine or solvent :

Traditionally, natural (organic) turpentine was used as the thinner or solvent, but has been replaced by several mineral based turpentine substitutes such as white spirit or "paint thinner".

6.16 TYPES OF VARNISHES

1. **Oil varnishes (or Oleoresinous varnishes) :** These are homogeneous solutions of one or more natural or synthetic resins in a drying oil and volatile solvent to which some driers are added. This type of varnish dries up slowly (drying period about 24 hours) by the evaporation of volatile solvent followed by oxidation and polymerisation of the drying oil. But the film produced is hard, quite lustrous and durable. e.g. Copal varnish (prepared by dissolving copal in linseed oil and turpentine). They are used for interior as well as exterior works.

2. **Spirit varnishes :** These varnishes are prepared by dissolving resin completely in volatile solvent like methylated spirit. In spirit varnishes instead of oil, spirit is used. Spirit varnishes dries up quite rapidly, but leaves behind a film which is brittle and so, has a tendency to crack or peel-off very soon. To avoid this some plasticizers are added. Moreover, the film is easily affected by weathering e.g., spirit varnishes are made by dissolving shellac (resin) in the spirit or alcohol.

6.17 USES OF VARNISHES

- Varnishes are used :
  1. For the protection of articles against atmospheric corrosion.
  2. As a brightening coat to the painted surface.
  3. For improving the appearance and intensifying the ornamental grains of wooden surfaces.

6.18 DISTINCTION BETWEEN PAINTS AND VARNISHES

Paints	Varnishes
1. Paint is a mechanical dispersion mixture of one or more pigments in a vehicle.	1. Varnish is a homogeneous colloidal dispersion solution of resins in oil or thinner or both.
2. Paint contains pigments.	2. Varnish do not contain pigments.
3. Paint obscure (i.e. hide) the surface on which it is applied.	3. Varnish do not obscure the surface on which it is applied.
4. Paint produces non-transparent (opaque) film.	4. It produces transparent film.
5. In paint, instead of oil, the resins cannot be used.	5. In varnish, instead of oil, the resins can be used.
6. Painted surface reflects heat and light nicely.	6. Varnished surfaces do not reflect heat and light.

6.19 JAPANS

- Japans are special type of pigmented varnishes, which are added to the paint to give it a good colour and lustre. Japans are of two types : (i) Printer's Japan and (ii) Decorative Japans (opaque varnish). The Printer's Japan consists of a resin dissolved in drying oil containing drier and thinner. They give more lustre to the paint. Decorative Japan consists of asphaltum or similar material and it provides colour and lustre to the paint. Decorative Japans are prepared by heating linseed oil and litharge (PbO) for about 5-6 hours at about 230°C, as a result of which a solid mass called lead oil is obtained. The lead oil is then mixed with asphaltum and a suitable thinner, such as kerosene to get decorative Japan or opaque varnish.

- Japans are used for painting bicycles, bed sheads and electrical devices.

6.20 ENAMELS

- "Enamel is a pigmented varnish (i.e., intimate dispersion of pigments in a varnish)." Enamels, therefore, give the combined advantages of a paint and a varnish. On drying they produce hard, lustrous and glossy finish. The properties of an enamel vary widely, depending largely on the nature of the varnish vehicle and resin.

- The drying of enamel (i) may occur at room temperature by oxidation (i.e. air drying enamel) and polymerisation, (ii) may be brought about at elevated temperatures either in the presence or absence of oxygen (baking enamel).

Constituents of enamels :

1. **Pigments :** Commonly used pigments are titanium dioxide (TiO_2) and calcium sulphate ($CaSO_4$) mixture. These are usually white, soft and having a fine texture. When coloured pigments are used, the enamels are called *Japans*. Black Japans (prepared by dissolving asphalt in linseed oil + turpentine or spirit) are very useful for painting metallic surfaces like bicycles, bed sheads and electrical devices. Metallic surface painted with Japan is baked at 210°C for 3-4 hours and then cooled. The oil gets oxidised by heat and the coating produced is highly resistant to chemicals and corrosion.

2.　**Vehicles :** Vehicles make enamels of glossy finish. The properties like viscosity, brushability, retention of gloss and colour are all affected by the choice of vehicle. The vehicle for enamel is either oleoresinous (oil + resin) or only resin. Natural resins like rosin and synthetic resins like alkyd resins are used as pure resin vehicle. In oleoresinous vehicles, synthetic resins like phenol aldehyde and oil (like linseed oil, soya bean oil or fish oil) are used.

3.　**Driers :** These are used only in case of oleoresinous enamels. The commonly used driers are naphthenates, oleates or resinates of Co, Zn, Mn, etc.

4.　**Thinners :** Thinners used are turpentine, xylol, acetone, etc.

6.21 LACQUERS

- "The dispersion of cellulose derivatives such as cellulose nitrate, resins and plasticizers in solvents and diluents is generally known as *lacquer*. Such a dispersion dries up by evaporation of its volatile constituents and provides hard, protective as well as decorative and water-proof film.

- Lacquer may be regarded as a modified form of spirit varnish that gives a transparent film. The lacquers are made in all colours that are commonly found in paints.

- The function of cellulose derivatives is to provide durability, hardness and water proofness, while that of natural and synthetic resins is to improve adhesion, water resistance capacity and retention of original gloss. Hence, resins are mainly used to improve the properties of film.

Constituents of Lacquer :

1.　**Cellulose derivatives :** Like cellulose acetate, cellulose nitrate and ethyl cellulose etc. give hardness, durability and water proofness to the film.

2.　**Resins :** Resins such as phenol formaldehyde, alkyd, copal etc. gives thickness, gloss, adhesion and water resistance to the film.

3.　**Solvents :** The solvents used are esters, ketones, ethers and alcohols (such as butyl alcohol, ethyl alcohol). The function of solvent is to dissolve the film forming material as well as to suspend the pigments. The choice of solvent depends upon the film forming material and the rate of evaporation of solvents, a defect known as "blushing" is developed. (Blushing means the formation of condensation of moisture upon lacquer film). Blushing can be avoided by using a solvent of high boiling point and hence slower rate of evaporation.

4.　**Plasticizers :** Plasticizers like castor oil, tributyl phosphate, chlorinated diphenyls etc. are added in lacquers to reduce brittleness, to improve adherence, durability and to decrease the inflammability of film.

5.　**Diluent :** Diluent is a non-solvent miscible with the solvent used. Diluents used are generally coal-tar products (such as tuluol, benzol) or petroleum products (such as petroleum naptha). The purpose of diluent added is to reduce the viscosity as well as cost of the lacquer.

Uses :

1.　Lacquers are generally used in the preparation of artificial leather and for giving finish to automobile parts (it resists abrasion, cracking and chalking).

2.　In interior decoration such as painting of wood work and furniture etc. (in which lacquers give a water-proof finish to the furniture).

3.　In the car industry, lacquers are used to give high gloss to car finishes.

Distinction between Oil varnish and Spirit varnish :

Oil varnish	Spirit varnish
1.　It is formed by dissolving resins in drying oil.	1.　It is formed by dissolving resins in methylated spirit.
2.　It dries up slowly.	2.　It dries up rapidly.
3.　It forms lustrous and durable film.	3.　It forms a brittle film.
4.　Plasticizers are not added to it.	4.　Plasticizers are added to it.
5.　Its film is not affected by weather.	5.　Its film is affected by weather.

Distinction between Enamels and Lacquers :

Enamels	Lacquers
1. It is a dispersion of pigment in varnish.	1. It is a dispersion solution of cellulose derivatives, resins and plasticizers in solvents and diluents.
2. It is known as pigmented varnish.	2. It is a modified spirit varnish.
3. Its film is hard, lustrous and glossy.	3. Its film is hard, transparent and water proof.

PRACTICE QUESTIONS

1. What is a paint ? For what purposes paints are used ?

2. What are the characteristics of good paint ?

3. Name the main constituents of paint. Give the functions of any one of them.

4. (a) Differentiate between varnish and lacquer.

   (b) Distinguish between paints and varnishes.

5. What are the demerits of distemper as compared to paint ?

6. What are the possible reasons for failure of paint films ? What are the remedies for it ?

7. What are the methods of application of paints ? Explain in brief.

8. Give any four characteristics of a good varnish or paint.

9. What is enamel ? Name the different constituents of enamels by giving one example of each.

10. (a) Give different causes of failure of paint film.

    (b) How the paint failure can be prevented ?

11. Define varnish. Give at least three points of difference between paint and varnish.

12. Define paint or varnish. Write four characteristics of a good paint.

13. Mention two purposes of varnishing or painting.

14. Define and explain lacquers and paints.

15. Name the different methods of applying a paint.

16. Define paint and varnishes.

17. Write types of constituents of varnishes.

18. Name the constituents of paint or varnish. Write any two functions of any two above constituents.

19. Write four characteristics of good varnish.

20. What are paints, varnishes, enamels and lacquers ? Name the different constituents of enamels and give one example of each.

21. What are emulsion paints ? State the constituents with examples. State the advantages and properties of emulsion paints.

22. Write a note on : (1) Cement paint, (2) Distempers, (3) Metal paints, (4) Luminous paints, (5) Exterior house paints, (6) Interior wall paints.

LUBRICANT AND LUBRICATION

7.1 LUBRICANT – DEFINITION (Nov. 18)

- Friction occurs in all types of machines, during motion of sliding or rolling surface of machines due to mutual rubbing of one part against the another. It causes :
  - (i) Wear and tear of two surfaces.
  - (ii) Loss of large amount of heat energy which decreases the efficiency of machines and
  - (iii) Deformation of moving parts of machines due to heat and thus shortening the life of machines.

- *To minimize such ill-effect of friction, a thin layer of suitable substance is applied between the moving parts of the machine. This is known as lubricant.*

- It is defined as *'any substance which is applied between two moving or sliding surfaces of machines to reduce the friction between the surfaces and minimize their wear and tear'. This process of application of lubricant between the moving or sliding surfaces of machines to reduce the friction between them is known as lubrication.*

7.1.1 Classification of Lubricants (Nov. 18)

- On the basis of physical state, lubricants are classified as solid, semi-solid and liquid lubricants.

(i) Solid lubricants :

Solid lubricants are used for certain systems where

  (1) The maintenance of liquid or semi-solid lubricant cannot be possible due to extreme operating conditions such as electric motors and generators.

  (2) The maintenance as lubricating oils or grease cannot be possible due to high operating temperature.

  (3) The machines are working on crude job under very high load and slow speed.

  (4) It is not easy to access the part which is to be lubricated.

Solid lubricants are used either in the form of powder or mixed with water or oil. Graphite and molybdenum disulphide are most common solid lubricants along with talc, mica, lead oxide, tungsten sulphide, aluminium stearates and aluminium palmitates etc.

Graphite is soft, soapy to touch, non-inflammable and not get oxidised in air, below 375°C is most popular solid lubricant which can be even used at high temperature in the absence of air. It is used in powder form or mixed with oil or water in presence of emulsifying agent like *tannin*. Graphite when dispersed in oil (known as oildag) is used in I.C. engine while Graphite dispersed in water (known as aquadag) is used in air compressors, food-stuff industry, heavy machines such as lathes etc. Graphite mixed with petroleum jelly and form graphite grease which are used as lubricant for railway track joints, chains, open gears, cast iron bearings etc. at high temperature.

(ii) Semi-solid lubricants :

The petroleum products such as grease, vaseline and waxes which are neither solid nor liquid at ordinary temperature are most important semi-solid lubricants. Out of which greases are most popular semi-solid lubricants used in case where

  (1) The splashes of oil spoil the whole machine.

  (2) The entry of dust or moisture spoil the lubricating action.

  (3) The oil cannot remain in place where it is subjected due to heavy load, slow speed and sudden jerk of parts of machines.

  (4) The bearing produces high temperature.

Grease is prepared from the mixture of petroleum, oil and soaps (i.e. sodium or potassium salt of higher salty acid) and it is classified into following types, depending on method of preparation and composition.

(1) Cup grease : It contains calcium oleate or stearate as stabilizer in emulsion of petroleum, oil and water. It is water resistant and used at low temperature below 80°C.

(2) Soda-base grease : It contains sodium soap in petroleum oil and is used at high temperature upto 175°C. However, it is not water resistant as sodium soap content is easily soluble in water. It is most suitable for *ball bearings*.

(3) Lithium base grease : It is prepared by mixing lithium soap with petroleum oil. It is water resistant and heat resistant used at very low temperature (15°C).

(4) Axle grease : It is a mixture of lime (or any metal hydroxide) and resin and fatty acids. This mixture is thoroughly mixed and allowed to stand. As a result grease floats as stiff mass and separated out. Axle grease is water resistant, used at high pressure and slow speed.

Other than grease, the mixture of mineral oils and solids like soap, graphite etc. are also used as semi-solid lubricants and are used in tractor rollers, mill axle boxes, wire ropes, machine bearings etc.

(iii) Liquid lubricants :

Liquid lubricants are used to large extent compared to solid and semi-solid lubricants because it not only reduces friction and wear between the two sliding or moving metallic surfaces but also act as cooling medium, sealing agent and good preventive from corrosion. Liquid lubricants are classified as

(1) Mineral oils : These are widely used as a lubricant because they are cheap, easily available in excess and are quite stable at working condition. These oils are obtained by fractional distillation of petroleum at a temperature of 400°C. The oils thus obtained are further subjected to number of processes, such as dewaxing, concentrated sulphuric acid treatment and solvent treatment etc. to remove wax and other impurities (e.g. asphalt) present in it. The presence of wax at low temperature makes the oil unfit for use as it can stiff the parts of machine and stop the motion of moving parts. The presence of impurities such as asphalt causes deposition of carbon and sludge formation at high temperature.

However, the mineral oils possess poor oiliness properties as they are quite thick in nature and hence are mixed with suitable animal or vegetable oils or compounds like oleic acid, stearic acid, etc. to increase its oiliness. Paraffin and naphthalene are mostly used as mineral oils.

(2) Vegetable oils : These are most commonly used lubricants before the discovery of petroleum. Now a days, they are mixed with mineral oils to improve its properties of oiliness. Some important vegetable oils are:

(i) Olive oil : It has colour from colourless to pale yellow and is obtained from olive tree. It is used to lubricate bearing and ordinary parts of machines which are subjected to low pressure and high speed.

(ii) Palm oil : It is a pleasant smell oil with pale yellow obtained from the kernels of palm fruit. It is used to lubricate delicate instruments like watches, clocks, scientific equipment etc.

(iii) Rape oil (or Colza oil) : It has characteristic smell and pale yellow colour and is obtained from colza plant. It is used to lubricate delicate apparatus and steam cylinders.

(iv) Castor oil : It is obtained from seeds of castor and is colourless to pale green colour with specific gravity of 0.16. It is used to lubricate rough parts of machines, vehicles and bearings which are subjected at high pressure.

(v) Hazel nut oil : It is used to lubricate watches, clocks and fine instrument is obtained from the hazel nut.

(3) Animal oils : They are resembles of vegetable oils obtained from animal sources and are mixed with mineral oils to improve its oiliness property. Some important animal oils are:

(i)   Neat foot oil : It is obtained by boiling fat of neat and water and is pale yellow colour oil with characteristic smell having specific gravity 0.915. To lubricate guns, sewing machines, watches, clocks and various parts of machines etc., neat foot oil is used as a lubricant.

(ii) Whale oil : It is thin, pale yellow oil obtained from fish whale by distillation process. It has specific gravity 0.88 and is used for light weighted machines.

(iii) Lard oil : The kidney, intestines and the fat of pigs produce colourless lard oil which is used to lubricate various types of machines.

(iv) Tallow oil : The fat of cattle produces transparent tallow oil which is mixed with mineral oils or without mixing also used as a lubricant.

(4)   Blended oils (or compounded oils) : The mixture of suitable oils are known as blended oils or compounded oils. When vegetable oils or animal oils are added to mineral oils, they are known as blended oils. Blended oil is used to reduce pour point, improve viscosity, increase oiliness, resist oxidation, reduce corrosion and improve colour of lubricant.

(5)   Synthetic oils : Synthetic oils used as lubricant in jet engines, rocket motors, submarines etc. are prepared chemically. Synthetic oils are used when mineral oils fail to work. However, these oils are very expensive.

7.1.2 Characteristics of Lubricants   (Nov. 18)

(1)   It avoided direct contact between the two rubbing surfaces and thus reduces surface deformation and wear and tear of the machine.

(2)   It acts as a coolant as it reduces loss of energy in the form of heat and thus enhance the efficiency of the machine.

(3)   Due to lubricant, the expansion of metal by frictional heat get reduced.

(4)   Lubricant increases smoothness property of moving parts of the machines.

(5)   It decreases the maintenance and running cost of the machine.

(6)   It reduces loss of power in internal combustion engine.

(7)   It acts as a good seal and thus prevent the leakage of gases at high pressure in internal combustion engine.

(8)   It also protects the parts of machine from destruction.

7.1.3 Functions of Lubricants   (Nov. 18)

Lubricants perform various functions. Some of these are :

- Reduces the loss of energy in the form of heat.
- Increases the efficiency of the machine.
- Prevents interlocking or inter joint welding at the surface asperities.
- Increases the smooth motion of moving parts.
- Reduces surface deformation, wear and tear.
- Reduces the expansion of metal by local frictional heat.
- Protects the material from corrosion.

7.2 LUBRICATION - MECHANISM OF LUBRICATION (TYPES OF LUBRICATION)   (Nov. 18)

It is of three types :

(1)   Fluid-film or thick-film or hydrodynamic lubrication.

(2)   Boundary or thin-film lubrication.

(3)   Extreme pressure lubrication.

1. Fluid-film (or thick film) lubrication :

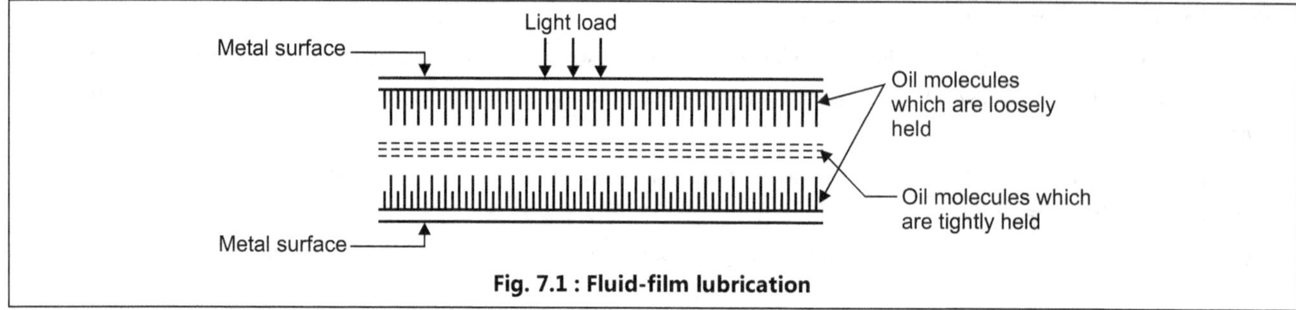

Fig. 7.1 : Fluid-film lubrication

- A suitable liquid lubricant of correct viscosity is placed between the moving or sliding surfaces of machines so that it fills or covers the irregularities of these surfaces and form a thick layer between them.

- This avoided direct contact between metallic surfaces and reduces wear of the machine. In this type, the chosen lubricant should be such that it should have very low frictional coefficient i.e. minimum viscosity under working condition and at the same time it remains in place and separates the surfaces.

- This is because the resistance of moving parts is only due to internal resistance between the particles of the lubricant moving over each other.

- This type of lubrication is applied to delicate instruments and light weighted high speed machines such as watches, clocks, sewing machines, guns, scientific instruments etc. Hydrocarbon oils are used as fluid-film lubricants.

2. Boundary (or thin film) lubrication :

- In certain cases as

    (a) When the machine starts from rest or

    (b) Speed is very low or

    (c) The load is very high or

    (d) Viscosity of oil is very low.

- The direct metal to metal contact is possible and continuous fluid-film lubricant cannot be maintained. In such cases, a thin layer of boundary lubricant oil is applied in space between the moving parts of machine which get adsorbed by physical force or chemical force or both.

- These adsorbed layers avoided direct contact of metal to metal surfaces as it is not easy to remove such layers.

- Vegetable and animal oils instead of mineral oils are used as boundary lubricant because they possess great property of adsorption due to their oiliness property.

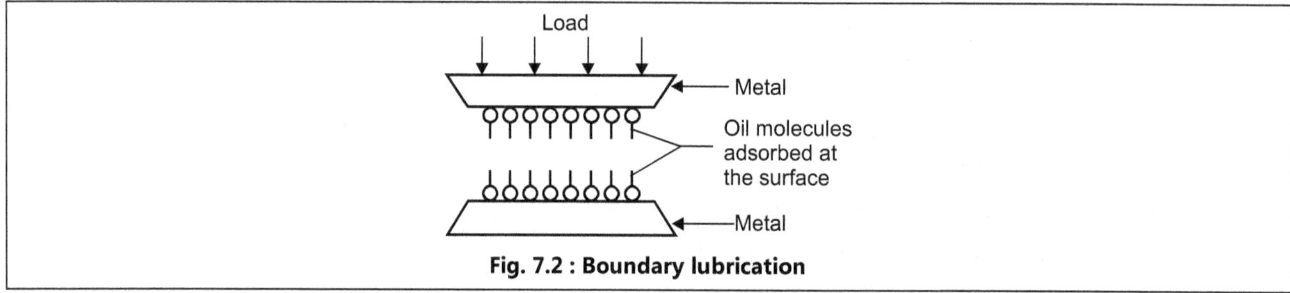

Fig. 7.2 : Boundary lubrication

- Solid lubricants like graphite and molybdenum disulphide act as excellent boundary lubricant either alone or as stable suspension in oil under the condition of very high loads and high temperature. This type of lubricant is provided to gears, tractors, rollers, etc.

3. Extreme pressure lubrication :

- It is done in such cases where the liquid lubricant fail to stick and may be decomposed and even vaporised, when sliding or moving surfaces of machines are working under high pressure, speed and at high temperature.

- In such extreme case where the liquid lubricant fail to stick and pressure lubricant i.e. a special additive known as *extreme pressure additive* is added to mineral oils is used which has ability to meet extreme pressure and high temperature condition.
- These additives are capable of withstanding high load and high temperature and therefore form durable film on the applied surface of metal.
- Organic compounds having certain active groups or radicals like chlorine, sulphur or phosphorus i.e. chlorinated esters, sulphurised oils and tricresyl phosphate are extreme pressure additives. Extreme pressure lubricant is required for cutting tools and steam turbines.

7.3 PHYSICAL CHARACTERISTICS AND TESTING OF LUBRICANTS

There are two tests : physical and chemical tests.

(A) Physical Properties with Experimental Determination :

(1) Viscosity, (2) Flash and Fire point, (3) Oiliness, (4) Cloud and Pour point, (5) Aniline point, (6) Emulsification, (7) Neutralization number, (8) Saponification number, (9) Oxidation stability.

1. Viscosity : The resistance offered by liquids to flow is called as viscosity. OR The force in dynes required to move 1 sq. cm of the liquid over another surface with a velocity of 1 cm/sec. The unit of viscosity is poise.

A high viscosity is thick and flows slowly. A low viscosity oil is thin and flows readily. Viscosity is determined by using Redwood viscometer.

Redwood viscometer : It is commonly used for determining viscosities of lubricating oils.

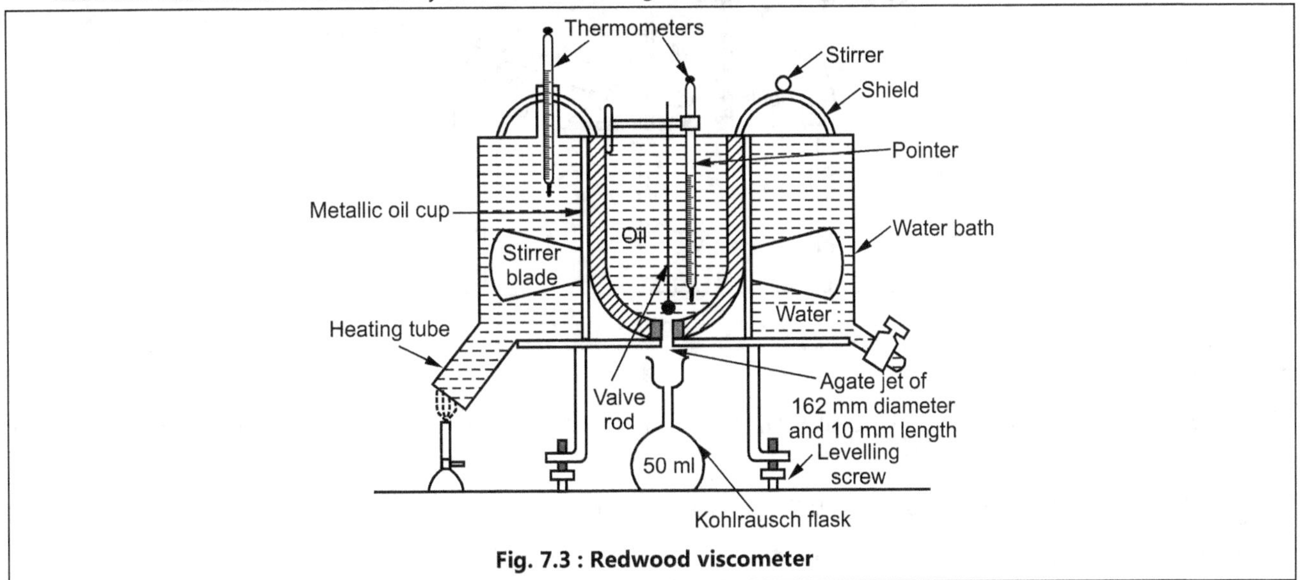

Fig. 7.3 : Redwood viscometer

It consists of following different parts :

(a) Oil cup : It is a cylindrical cup made up of brass and silver plated from inside having 90 mm height and 46.5 mm in diameter. The cup is open at the upper end. At the bottom of the cup, a jet of diameter 1.62 mm and internal length of 10 mm is fitted.

(b) Heating bath : The oil cup is surrounded by a cylindrical copper vessel containing water. It is provided with tap for filling and emptying the water.

(c) Flask : It is specially shaped for receiving the oil from the jet outlet. Its capacity is 50 ml upto the mark in its neck.

(d) Stirrer : A stirrer with four blades outside the oil cup in water bath is provided to maintain uniform temperature in the bath.

Working : The oil cup is cleaned thoroughly. The jet is kept closed with ball of the valve rod at it. Oil under test is filled in clean cup upto the pointer level. The bath is filled with water and the side tube is heated with constant stirring. When the desired temperature is obtained, the ball is lifted and the stop watch is started at the same instant. When the oil reaches at the mark on the neck of the flask, the stop watch is stopped and the valve is closed to prevent the overflow of oil. Thus note the time in seconds required for 50 ml of the oil.

2.   **Viscosity index :** The rate of change of viscosity of liquid with temperature is known as viscosity index.

The viscosity of liquid varies with the change of temperature. A good lubricant is that whose viscosity does not vary much or differ with the temperature.

3.   **Flash point and Fire point :** *"Flash point is the lowest temperature at which the oil gives off enough vapours when small flame is brought near it"*. In this case, fire only takes place but does not burn continuously.

"Fire point is the lowest temperature at which the oil burns continuously for at least five seconds when a small flame is brought near to it". In many of the cases, fire points are 5 to 40°C higher than its flash point. A lubricating oil having low flash and fire point is not safe because due to rubbing of machine parts in contact, some heat is produced even if the lubricant has been applied in between as a result, the temperature is increased.

The flash and fire points are usually determined by using Pensky-Marten's apparatus.

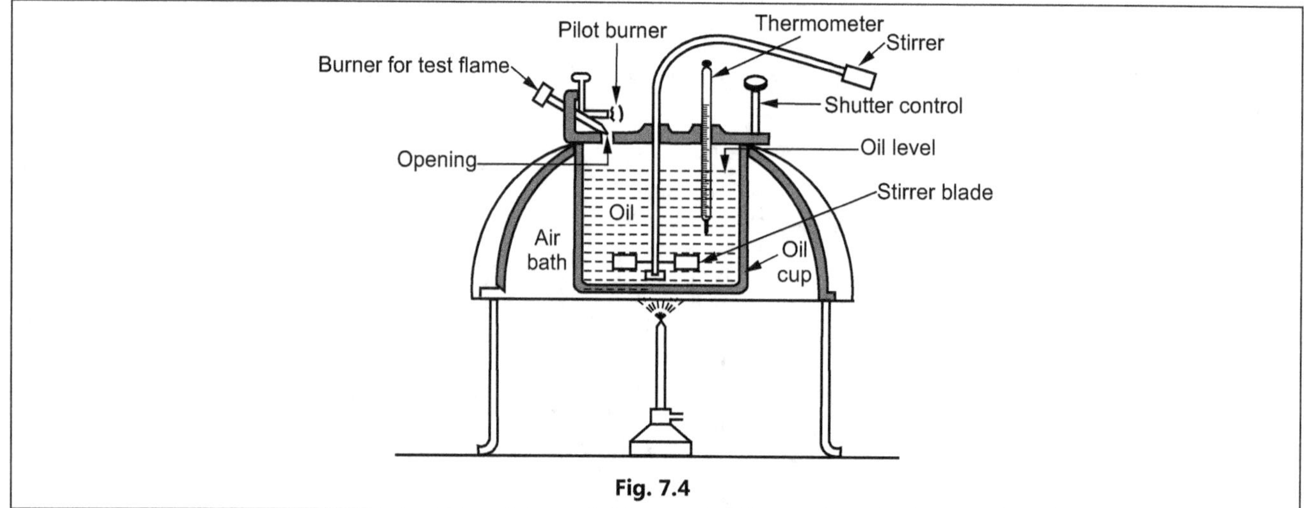

Fig. 7.4

Construction :

(i)   **Oil cup :** The diameter of cup is about 5 cm and 5.5 cm deep. The cup lid is provided with four openings of standard dimensions.

(ii) **Air bath :** The oil cup is supported by its flange over in air bath which is heated by gas burner.

(iii) **Piolet burner :** As test flame is introduced in the cup, it gets extinguished but when the test flame is returned to its original position, it is relighted by the piolet burner. i.e. automatically lighted flame due to piolet burner.

(iv) **Flame exposure device :** It is a small flame and this is connected to the shutter by a lever mechanism.

Working : The oil to be tested is filled upto the mark in the oil cup. The air bath is heated by the burner. The stirrer is kept working continuously at the rate of 1 to 2 revolutions per second. The heating is done in such a way that the temperature rises by 5°C per minute. Flame is introduced through the opening at every 1°C rise of temperature.

The temperature at which a distinct flash appears inside the cup is recorded as flash point. The heating is continued and the temperature at which oil ignites and continues to burn for at least 5 seconds is recorded as the fire point.

4.   **Cloud point and Pour point :** Cloud point indicates the temperature at which the oil becomes cloudy in appearance. Pour point of a liquid is the temperature at which the liquid ceases to flow on cooling.

Cloud point and pour point indicate the suitability of lubricants used in cold condition.

The cloud point and pour point of lubricants can be determined with the help of pour point apparatus as shown in Fig. 7.5.

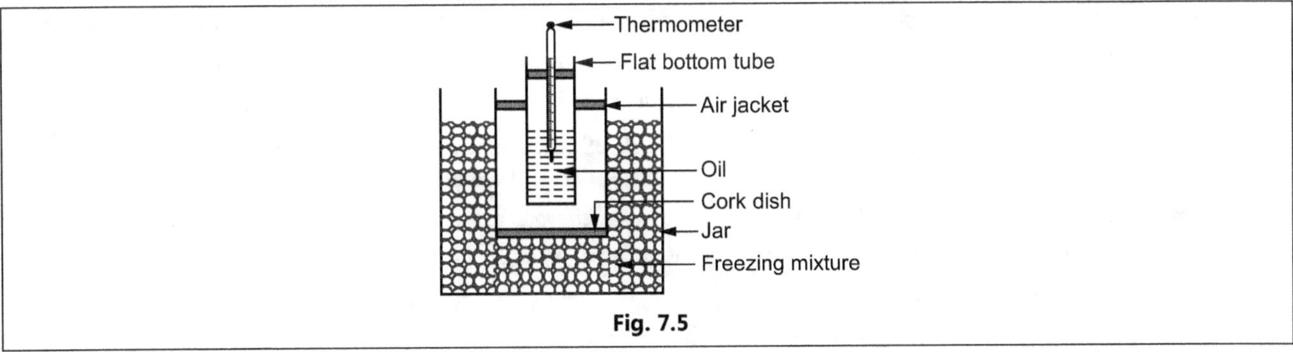

Fig. 7.5

The apparatus consists of a flat bottom tube about 3 cm in diameter and 12 cm high. It is fitted in air jacket surrounded by the freezing mixture. The tube is filled with oil. A thermometer through cork is introduced in the oil.

As the cooling proceeds, the temperature of the oils falls down. *The temperature at which oil becomes cloudy or hazy in appearance is recorded as the cloud point.*

After this the cooling is continued and the test tube is withdrawn after every 3°C fall of temperature and tilted to see the oil flows or not. The temperature at which oil does not flow in the test tube even when kept horizontal for 5 seconds is recorded as the pour point.

5. Oiliness : It is defined as the *power of an oil to maintain a continuous film under pressure while used as a lubricant.*

It is the property of an oil by which it sticks on the surface of machine parts working under high pressure and high temperature. Mineral oils have comparatively low oiliness than the vegetable oils. The oiliness of the mineral acid can be increased by the addition of fatty acids such as stearic acid or oleic acid.

(B) Chemical properties with determination :

1. Aniline point : Aromatic hydrocarbons have a tendency to dissolve natural as well as synthetic rubbers. The percentage of hydrocarbons is decided on the basis of aniline point. Higher the aniline point, lower is the percentage of aromatic hydrocarbons. The tendency of the lubricant to mix with aniline is also expressed in terms of aniline point. *Aniline point is the minimum equilibrium solution temperature for equal volumes of aniline and the lubricant sample.* **OR** *It is the lowest temperature at which oil is completely miscible with an equal volume of freshly distilled aniline.*

Procedure : Aniline point is determined by taking equal volumes of lubricating oil sample and aniline in a dried test tube. The tube is then heated in paraffin bath with constant stirring until homogeneous solution is obtained. Then stop the heating and tube is allowed to cool at a rate of 1°C per minute with constant stirring.

The temperature at which the two phases (oil and aniline) separate out is recorded as the aniline point.

2. Neutralization number : The acidity and alkalinity of lubricating oil is determined in terms of neutralization number.

The acidity of an oil is expressed in acid value or acid number and is defined as the *number of milligrams of KOH required to neutralize the free acid present in 1 gm of an oil.* The acid value of lubricating oil is always less than one.

Determination of Neutralization number or Acid value : A known quantity of oil is taken in a titration flask. Add 50 ml alcohol + 50 ml water, titrate it against 0.1 N KOH solution by using phenolphthalein as an indicator. The end point is pink to colourless.

$$\text{Acid number} = \frac{\text{Number of cc of 0.1 N KOH} \times 5.6}{\text{Weight of oil taken}}$$

3. Saponification number : It is defined as the *number of milligrams of KOH required to saponify 1 gm of an oil.*

Determination of saponification number : Add 25 ml of 0.5 N alcoholic KOH to the oil sample. Reflux it on water bath for 1 hour, so that saponification will be complete. Then titrate it against 0.5 N HCl using phenolphthalein as an indicator, simultaneously reflux 25 ml 0.5 N alcoholic KOH on water bath for 1 hour and titrate it against 0.5 N HCl. The difference in both the titrations gives the quantity of 0.5 N KOH required for saponification.

$$\text{Saponification value} = \frac{\text{ml of KOH required} \times \text{Molecular weight of KOH} \times \text{Normality}}{\text{Weight of oil sample}}$$

4. Emulsification : Emulsification is the *property of oils to mix with water forming an intimate mixture.* It have a tendency to pick up dirt, dust particles and foreign matter which spoil the lubricating action.

A good lubricant should not form emulsions or emulsions should break off quickly. The time in seconds in which oil and water separate out in a distinct layer is called steam emulsion number (S.E.N.).

5. Volatility : Volatility of a liquid is its tendency to vaporize with the increase of temperature. If a lubricant is highly volatile, it will vaporize rapidly at low temperature. Hence extra quality of lubricant is required. The lubrication process will be costly. A good lubricant should have low volatility.

7.4 GENERAL CRITERIA FOR SELECTION OF LUBRICANTS

Selection of Lubricants : Lubricants reduce wear and tear and energy loss due to friction. It provides a seal around the piston ring in I.C. engine. The properties of the lubricant should not change during the service period.

Properties of lubricants :

(1) By knowing the properties of lubricant and the condition for which lubrication required, suitable lubricant must be selected so that its properties should not change under service conditions.

(2) The selected lubricant should neither be very thick nor very thin in nature.

(3) In moving parts of the machine under pressure, selected lubricant should form continuous layer.

(4) It should not prevent the movement of various parts of the machine by getting stuck to it.

For various types of machines, depending on their service conditions, different lubricants are used. They are illustrated as follows.

(1) Lubricants for delicate instruments : For delicate instruments, thin vegetable oils or animal oils such as palm oil, hazel nut oil, neat foot oil etc. are used as lubricants because these instruments are exposed only at low temperature and light load. Watches, clocks, sewing machines, scientific instruments etc. are delicate instruments.

(2) Lubricants for very high pressure and low speed machines : Grease or thick oil film cannot be maintained for machines in which very high pressure and low speeds are employed. Hence for such machines like tractor rollers, lathes, railway track joints, concrete mixtures etc. graphite, mica, soap-stone, molybdenum disulphide like solid lubricants are used.

(3) Lubricants for high pressure and low speed machines : For high speed and low pressure machines such as rail axle boxes, wire ropes etc. grease or thick oils or thick blended oils are used as lubricants because there is a chance for thin oils to get squeezed out due to high pressure.

(4) Lubricants for gears : Lubricants which possess good oiliness, high oxidation stability and high load bearing capacity are selected as lubricants for gears as it is subjected to extreme pressure. Hence thick mineral oils containing extreme pressure additives like metallic soap, sulphur or chlorine or phosphorus compound etc. are used as lubricants for gears.

(5) Lubricants for cutting tools : The tools which are used for cutting, turning, boring, drilling etc. produce high friction and generate large amount of heat during its operation. Hence lubricants required for such process should be such that it cool the tool and reduce the friction. Therefore for heavy cutting tools, mineral oils of low viscosity containing additives like fatty acids, sulphurised fatty acid and chlorinated compounds are used as lubricants. For *light cutting tools*, oil emulsions are used as lubricants because it acts as an efficient cooling media.

(6) Lubricants for Internal Combustion engine : For internal combustion engine (I.C. engine), lubricants should possess high viscosity index and high thermal stability is required as they are exposed to very high temperature. Hence in I.C. engine, mineral oils containing additives which impart high viscosity index are used as lubricants.

(7) Lubricants for steam turbines : The lubricants which impart oxidation stability, corrosion resistance and demulsification properties such as blended oils containing additives are used in steam turbines. Because lubricants are exposed to high temperature and oxidising conditions and it comes in contact with water due to leakage.

(8) Lubricants for steam engine cylinder : The mineral oils mixed with vegetable oils or mineral oils having high viscosity are used as lubricants for steam engine cylinder.

(9) Selection of lubricants for electrical transformers : In case of electrical transformers, the role of lubricating oil is to insulate the windings and to carry away the heat generated, when transformer is on load. The oil must possess good dielectric properties and since effective heat transfer depends on continuous circulation, it must be of low viscosity. When transformer is no load, the exposure of air, high temperature, electrical stress and catalytic influence of copper windings, etc. cause various chemical changes in oils leading to the formation of acid as well as sludge. The acid thus formed may attack the metallic parts of the transformer like copper wire, transformer tank etc. Hence to minimize the formation of acid and sludge, highly refined mineral oils having insulating properties, strong oxidation resistance and high chemical stability are used for electrical transformers.

(10) Selection of lubricants for hydraulic systems :

(a) The important characteristics required are proper viscosity.

(b) The oil should have high viscosity index.

(c) It should have deemulsification character.

(d) It should have good oxidation resistance along with rust preventing character.

(e) It should have anti-wear characteristics.

(f) The viscosities of oils required in hydraulic systems are 150, 210, 310 and 400 SUS at 100°F for highest grades.

(11) Selection of lubricants for refrigeration :

(a) Oils should have low pour and cloud point. Also low flow point.

(b) Naphthenic base oils, hence, can be employed in refrigeration.

(c) The viscosities should be in 85, 160, 200 and 325 SUS at 100°F.

(d) Oil must withstand in determined phase at low temperature.

7.5 CHARACTERISTICS OF TRANSFORMER OIL

- Transformer oil is basically an insulating oil. Insulating oils or transformer oils are mineral electrical insulating oils processed from fractional distillation and treatment of crude petroleum.

  An insulating oil must have the following properties :

  1. Excellent dielectric properties resulting in minimum power loss.

  2. High resistivity leading to better insulation values between windings.

  3. High flash point and thermal stability facilitating reduction in evaporation losses.

  4. Long life performance and excellent ageing characteristics, even under severe electrical stress.

  5. Virtually devoid of corrosive sulphur, leading to enhanced stability and protection against corrosion.

  6. Wider operating temperature range, resulting from higher flash and lower pour point.

  7. Available in a wide range of grades to fit every requirement and meeting every possible specification.

- Insulating oil when used in a transformer is popularly known as transformer oil. The oil in transformer serves the following purposes :

  1. Transformer oil provides the liquid insulation in between the windings and other conductive parts of transformer.

  2. It acts as a cooling medium.

  3. It preserves the core and the winding of the transformer as the transformer core and winding remain immersed in the oil.

  4. The most important role played by the transformer oil is to protect the interaction of atmospheric oxygen and the cellulose made paper insulation of the winding. In this way, transformer oil prevents the oxidation of cellulose made paper insulation of the winding.

- Thus we observe that transformer oil has great importance and it does not only act as a cooling medium rather it does many other things to keep transformer healthy.

PRACTICE QUESTIONS

1. Define lubricant and lubrication. Give characteristics of a lubricant.
2. Explain different types of mechanism of lubrication.
3. Explain determination of viscosity of oil by Redwood viscometer with suitable diagram.
4. Explain determination of flash and fire point of oil by Pensky-Marten's apparatus with diagram.
5. Explain cloud and pour point with suitable sketch.
6. Define and explain determination of acid value of an oil.
7. Write in short saponification and emulsification.
8. Give the general criteria for the selection of lubricant for :
   - (i) I.C. engines,
   - (ii) Refrigeration system,
   - (iii) Gears,
   - (iv) Cutting tools,
   - (v) Transformers.
9. Define :
   - (i) Lubrication,
   - (ii) Flash point,
   - (iii) Fire point,
   - (iv) Cloud point,
   - (v) Pour point,
   - (vi) Viscosity index,
   - (vii) Oiliness,
   - (viii) Acid value.
10. Write a note on semi-solid lubricants.
11. Give the characteristics of transformer oil.

1. **Choose one correct alternative from the four alternatives :** (2 × 8 = 16)

(i) Which solution maintains its pH value ?

   (a) dil. HCl

   (b) dil. H_2SO_4

   (c) dil. NaOH

   (d) Buffer solution.

(ii) The purest form of iron is ......

   (a) Pig iron

   (b) Wrought iron

   (c) Red iron

   (d) Steel.

(iiii) Anode mud is the by-product of ......

   (a) Liquation

   (b) Electrolytic refining

   (c) Smelting

   (d) Calcination.

(iv) Main components of Tinman's solder is ......

   (a) Sn, Pb

   (b) Sn, Cu

   (c) Bi, Pb

   (d) Cu, Al.

(v) Which one is good adhesive ?

   (a) Epoxy resin

   (b) Ceramics

   (c) Bauxite

   (d) Rose metal.

(vi) A substance present in temporary hard water is ......

   (a) Magnesium chloride

   (b) Magnesium bicarbonates

   (c) Magnesium sulphate

   (d) All of the above.

(vii) EDTA method is used for determining ......

   (a) Temporary hardness

   (b) Permanent hardness

   (c) Temporary and Permanent hardness

   (d) Alkalinity.

(viii) Coating of Tin to Iron is called :

   (a) Priming, (b) Painting, (c) Tinning, (d) Spraying.

2. **(a)** Write the name of important ores of Iron and Copper. Explain the smelting process of iron ore inside the blast furnace with reactions. **(8)**

Ans. Refer to Section 2.2.2 on page 2.8. Also refer to Section 2.2.3 on page 2.9.

(b) What is ferrous alloy ? Write a note on types of steel. **(8)**

Ans. Refer to Section 2.4 on page 2.28. Also refer to Section 2.4.1 on page 2.28.

3. **(a)** Define refractories and composite materials. Write their properties and applications with example. **(8)**

Ans. Refer to Sections 3.2.1, 3.2.5 and 3.2.6 on page 3.3 to 3.5. Also refer to Sections 3.3.1, 3.3.2 on page 3.7.

(b) What do you mean by hard and soft water ? Explain any one method for removal of hardness from water. **(8)**

Ans. Refer to Section 4.4 on page 4.2. Also refer to Section 4.5 on page 4.2.

4. **(a)** What is corrosion ? Explain types of corrosion and the factors affecting corrosion. **(8)**

Ans. Refer to Sections 5.1 and 5.2 on page 5.1. Also refer to Section 5.3 on page 5.10.

(b) What is paint ? Write a note on heat resistant paint and cellulose paint. **(8)**

Ans. Refer to Section 6.1 on page 6.1. Also refer to Section 6.5 on page 6.3.

5. **(a)** What is lubricant and lubrication ? Explain types of lubricant and function of lubricant. **(8)**

Ans. Refer to Sections 7.1, 7.1.1, 7.1.2, 7.1.3 and 7.2 on page 7.1 to 7.3.

(b) What do you mean by protective coatings on metal surface ? Write a note on galvanising and sherardizing. **(8)**

Ans. Refer to Section 5.5 on page 5.14.

6. **Write notes on any four :** **(4 × 4 = 16)**

(a) Buffer solution.

Ans. Refer to Section 1.13 on page 1.14.

(b) Adhesives.

Ans. Refer to Section 3.4.1 on page 3.8.

(c) Varnishes.

Ans. Refer to Section 6.13 on page 6.11.

(d) Viscosity and Viscosity index.

(e) Scale and Sludge formation.

Ans. Refer to Section 4.10 on page 4.6.

(f) Electroplating.

Ans. Refer to Section 5.4 on page 5.14.

❏❏❏